I0465531

NOT MEASUREMENT
SENSITIVE

NASA TECHNICAL HANDBOOK

National Aeronautics and Space Administration

NASA-HDBK-8709.22
with Change 4

Approved: 2018-03-08
Superseding NASA-STD-8709.22

SAFETY AND MISSION ASSURANCE ACRONYMS, ABBREVIATIONS, AND DEFINITIONS

DOCUMENT HISTORY LOG

Status	Document Revision	Approval Date	Description
Baseline		03/08/2018	Changed document from a Technical Standard to a Technical Handbook. Removed definitions not used in Agency SMA directives and standards. Added and updated all definitions used in Agency SMA directives and standards. Revised the document format.
Change	1	03/20/2018	Updated for recently published NPR 8715.5, NASA-STD-8719.25 and NASA-STD-8739.12.
Change	2	04/03/2018	Updated to account for definitions added and edited in recently published NASA-STD-8719.24 Annex.
Change	3	06/18/2018	Updated to account for added, edited, and deleted acronyms (~8) and definitions (~23) in the recently published NASA-STD-8719.12A.
Change	4	11/28/2018	Updated to account for added, edited, and deleted acronyms (~22) and definitions (~68) in the recently published NPR 8715.6B with Change 1, NPR 8735.1D, NASA-STD-8719.9B, and NASA-STD-8719.17C with Change 2.

FOREWORD

The National Aeronautics and Space Administration published this Technical Handbook as a compilation of definitions of acronyms, abbreviations, and terms used in NASA safety and mission assurances directives and standards. It is provided as a convenience, and as a means of identifying differences and commonalities in the use of these terms. Please refer to the acronyms and definitions in each individual document for the most up to date definitions.

Requests for information, corrections, or additions to this handbook should be submitted to the NASA, Office of Safety and Mission Assurance (OSMA), by email to Agency-SMA-Policy-Feedback@mail.nasa.gov or via the "Submit Feedback" link at https://standards.nasa.gov.

Terrence W. Wilcutt
NASA Chief, Safety and Mission Assurance

3/8/2018
Approval Date

TABLE OF CONTENTS

1. SCOPE

1.1 Purpose

This NASA Technical Handbook compiles into a single volume safety, reliability, maintainability, and quality assurance and risk management terms defined and used in NASA safety and mission assurance directives and standards. The purpose of this handbook is to support effective communication within NASA and with its contractors. The definitions in this handbook are updated when the definition of the acronym or term is updated in the originating document.

1.2 Applicability

This NASA Technical Handbook is a reference and is provided for convenience only. Please refer to the originating document for the most up to date definitions.

2. NASA ORIGINATING DOCUMENTS

2.1 General

2.1.1 The documents listed in this section define and cite the abbreviations, acronyms, and definitions provided in this handbook.

2.1.2 The following documents are accessible via the NASA Online Directives Information System (NODIS) at http://nodis3.gsfc.nasa.gov/ or the NASA Technical Standards System at http://standards.nasa.gov.

NPR 8000.4	Agency Risk Management Procedural Requirements
NPR 8621.1	NASA Procedural Requirements for Mishap and Close Call Reporting, Investigating, and Recordkeeping
NPR 8705.2	Human-Rating Requirements for Space Systems
NPR 8705.4	Risk Classification for NASA Payloads
NPR 8705.5	Technical Probabilistic Risk Assessment (PRA) Procedures for Safety and Mission Success for NASA Programs and Projects
NPR 8705.6	Safety and Mission Assurance (SMA) Audits, Reviews, and Assessments
NPR 8715.3	NASA General Safety Program Requirements
NPR 8715.5	Range Flight Safety Program

NPR 8715.6	NASA Procedural Requirements for Limiting Orbital Debris and Evaluating the Meteoroid and Orbital Debris Environments
NPR 8715.7	Expendable Launch Vehicle (ELV) Payload Safety Program
NPR 8735.1	Procedures for Exchanging Parts, Materials, Software, and Safety Problem Data Utilizing the Government-Industry Data Exchange Program (GIDEP) and NASA Advisories
NPR 8735.2	Management of Government Quality Assurance Functions for NASA Contracts
NASA-STD-6008	NASA Fastener Procurement Receiving Inspection & Storage Practices for Spaceflight Hardware
NASA-STD-8709.20	Management of Safety and Mission Assurance Technical Authority (SMA TA) Requirements
NASA-STD-8719.7	Facility System Safety Guidebook
NASA-STD-8719.9	Lifting Standard
NASA-STD-8719.11	Safety Standard for Fire Protection
NASA-STD-8719.12	Safety Standard for Explosives, Propellants, and Pyrotechnics
NASA-STD-8719.13	Software Safety Standard
NASA-STD-8719.14	Process for Limiting Orbital Debris
NASA-STD-8719.17	NASA Requirements for Ground-Based Pressure Vessels and Pressurized Systems (PVS)
NASA-STD-8719.24	NASA Expendable Launch Vehicle Payload Safety Requirements
NASA-STD-8719.24 Annex	NASA Expendable Launch Vehicle Payload Safety Requirements
NASA-STD-8719.25	Range Flight Safety Requirements
NASA-STD-8729.1	NASA Reliability and Maintainability (R&M) Standard for Spaceflight and Support Systems
NASA-STD-8739.1	Workmanship Standard for Polymeric Application on Electronic Assemblies

NASA-STD-8739.4	Workmanship Standard for Crimping, Interconnecting Cables, Harnesses, and Wiring
NASA-STD-8739.5	Workmanship Standard for Fiber Optic Terminations, Cable Assemblies, and Installation
NASA-STD-8739.6	Implementation Requirements for NASA Workmanship Standards
NASA-STD-8739.8	Software Assurance Standard
NASA-STD-8739.9	Software Formal Inspections Standard
NASA-STD-8739.10	Electrical, Electronic, and Electromechanical (EEE) Parts Assurance Standard
NASA-STD-8739.12	Metrology and Calibration

3. ACRONYMS AND ABBREVIATIONS

Abbreviations for units of measure used in the citing documents are not included here.

Acronym	Term
°C	Degree Celsius
°C	Degree Centigrade
°F	Degree Fahrenheit
30 SW/SE	30Th Space Wing/Chief of Safety
30 WS	30Th Weather Squadron
A&E	Architect Engineering
A-50	Aerozine 50 (Udmh)
AA	Associate Administrator
AA/HEOMD	Associate Administrator, Human Exploration and Operations Mission Directorate
AA/OCOM	Assistant Administrator, Office of Communications
AA/OIIR	Associate Administrator, Office of International and Interagency Relations
AA/OIIR	Assistant Administrator, Office of International and Interagency Relations
AAO	Audits and Assessments Office
AC	Alternate Current
AC	Alternating Current
ac	Alternating Current
ACGIH	American Conference of Governmental Industrial Hygienists
ACGIH	American Councils of Governmental Industrial Hygienists
ACI	American Concrete Institute
ADA	Americans With Disabilities Act
AF	Air Force
AFB	Air Force Base
AFFF	Aqueous Film Forming Foam
AFI	Air Force Instruction
AFJMAN	Air Force Joint Manual
AFMAN	Air Force Manual

Acronym	Term
AFOSH	Air Force Occupational Safety and Health
AFPD	Air Force Policy Directive
AFSPC	Air Force Space Command
AFSPCMAN	Air Force Space Command Manual
AFSS	Autonomous Flight Safety System
AGA	American Gas Association
AGE	Aerospace Ground Equipment
Agency Team	NASA ELV Payload Safety Agency Team
AGMA	American Gear Manufacturers Association
AGS	Aboveground Sites
AHJ	Authority Having Jurisdiction
AIA	Aerospace Industries Association
AIP	Acquisition Integrity Program
AISC	American Institute of Steel Construction
AMD	Aircraft Management Division
ANSI	American National Standards Institute
AO	Appointing Official
AOA	Angle of Attack
AoA	Analysis of Alternatives
AO-HRR	American Optical Hardy-Rand-Rittler
API	American Petroleum Institute
ARFF	Aircraft Rescue and Fire-Fighting
ASAP	Aerospace Safety Advisory Panel
ASHRAE	American Society of Heating, Refrigeration, and Air Conditioning Engineers
ASIC	Application-Specific Integrated Circuit
ASIC	Application Specific Integrated Circuit

Acronym	Term
ASM	Acquisition Strategy Meeting
ASME	American Society of Mechanical Engineers
ASNT	American Society for Nondestructive Testing
ASQ	American Society for Quality
ASTM	American Society for Testing and Materials
AVE	Aerospace Vehicle Equipment
AWG	American Wire Gage
BDA	Blast Danger Area
BPVC	Boiler and Pressure Vessel Code
BTHLD	Below-The-Hook-Lifting-Device
CAAS	Contract Administration and Audit Services
CAD	Cartridge-Activated Device
CAGE	Commercial and Government Entity
Cal/OSHA	California Occupational Safety and Health Administration
CAL-OSHA	California Occupational Safety and Health Administration
CAP	Corrective Action Plan
CARA	Conjunction Assessment Risk Analysis
CASE	Computer-Aided Software Engineering
CCAFS	Cape Canaveral Air Force Station
CCCV	Constant Current Constant Voltage
CCP	Counterfeit Control Plan
CD	Center Director
CD	Compact Disc
CD ROM	Compact Disc Read Only Memory
CDR	Critical Design Review
CE	Complex Electronics
CEN	European Committee for Standardization
CERR	Critical Events Readiness Review

Acronym	Term
CFR	Code of Federal Regulations
CGA	Compressed Gas Association
Chief/OSMA	Chief, Office of Safety and Mission Assurance
CI	Configuration Item
CIL	Critical Item List
CIS	Certified IPC® Application Specialist
CIT	Certified IPC® Trainer
CMAA	Crane Manufacturers Association of America
CMM®	Capability Maturity Model
CMMI	Capability Maturity Model Integration
CMM-SW	Capability Maturity Model - Software
CMO	Configuration Management Organization
CMOR	Canadian Meteor Orbit Radar
CMS	Contingency Management System
CMS	Configuration Management System
CO	Contracting Officer
COA	Certificate of Authorization or Waiver
COB	Chip on Board
COC	Certificate of Conformance
CoF	Construction of Facilities
CoFR	Certification of Flight Readiness
COLA	Collision Avoidance
COPV	Composite Overwrapped Pressure Vessel
COTR	Contracting Officers Technical Representative
COTS	Commercial-off-the-Shelf
COTS	Commercially Available off-the-Shelf
COTS	Commercial Off-The-Shelf
COTS	Commercial off the Shelf
COTS	Computer off-The-Shelf

Acronym	Term
COTS	Commercial off-The-Shelf Software
CPIA	Chemical Propulsion Information Agency
CPLD	Complex Programmable Logic Devices
CRF	Critical Radiant Flux
CRFSL	Center Range Flight Safety Lead
CRM	Continuous Risk Management
CSCI	Computer Software Configuration Item
CSO	Chief Safety and Mission Assurance Officer
CTE	Coefficient of Thermal Expansion
CVCM	Collected Volatile Condensable Material
CVD	Chemical Vapor Deposition
CVT	Certification Validation Testing
CW	Continuous Wave
DAS	Debris Assessment Software
DASHO	Designated Agency Safety and Health Official
DC	Direct Current
dc	Direct Current
DCMA	Defense Contract Management Agency
DCR	Design Certification Review
DDD	Displacement Damage Dose
DDESB	Department of Defense Explosives Safety Board
DFMR	Design for Minimum Risk
DFO	Distant Focusing Overpressure
DIN	Deutsches Institut für Normung
DIP	Dual-In-Line Package
DMSMS	Diminishing Manufacturing Sources and Material Shortages
DoD	Department of Defense
DOE	Department of Energy

Acronym	Term
DOT	Department of Transportation
DPA	Destructive Physical Analysis
DR	Decommissioning Review
DR	Discrepancy Report
DSN	Deep Space Network
DWV	Dielectric Withstanding Voltage
EAR	Export Administration Regulations
EAV	Experimental Aeronautical Vehicle
EBW	Exploding Bridgewire
EBW-FU	Exploding Bridgewire Firing Unit
E_C	Expectation of Casualty
ECM	Earth-Covered Magazine
ECP	Engineering Change Proposal
ED/OHO	Executive Director, Office of Headquarters Operations
EED	Electro-Explosive Device
EED	Electroexplosive Device
EEE	Electrical, Electronic, and Electromechanical
EELP	Explosives, Energetic Liquids, and Pyrotechnics
EGSE	Electrical and Electronic Ground Support Equipment
EIA	Electronic Industries Association
EID	Electrically Initiated Device
ELCG	Energetic Liquid Compatibility Group
ELDRS	Enhanced Low Dose Rate Sensitivity
ELS	Equivalent Level of Safety
ELV	Expendable Launch Vehicle
EMC	Electromagnetic Compatibility
EMD	Environmental Management Division
EME	Electromagnetic Energy
EMI	Electromagnetic Interference
EMR	Electromagnetic Radiation
EMS	Emergency Medical Service

Acronym	Term
E-NMTTC	Eastern NASA Manufacturing Technology Transfer Center
EOD	Explosive ordnance Disposal
EOL	End of Life
EOM	End of Mission
EOMP	End of Mission Plan
EOPR	End of Period Reliability
EPA	Electrostatic Protected Area
EPA	Environmental Protection Agency
EPARTS	Electronic Parts Applications Reporting and Tracking System
EPMCP	EEE Parts Management and Control Plan
EPROM	Erasable Programmable Read Only Memory
ER	Eastern Range
ERM	Enterprise Risk Management
ERP	Effective Radiated Power
ERP	Emergency Response Plan
ERPG	Emergency Response Planning Guidelines
ES	Exposed Site
ESA	European Space Agency
ESD	Electrostatic Discharge
ESDS	Electrostatic Discharge Sensitive
ESO	Explosive Safety Officer
ESO	Explosives Safety Officer
ESQD	Explosive Safety Quantity Distance
ESS	Environmental Stress Screening
ESSP	Earth Systems Science Pathfinder
E-Stop	Emergency Stop
ETA	Explosive Transfer Assembly
ETBA	Energy Trace Barrier Analysis
ETS	Explosive Transfer System
EVA	Extravehicular Activity
FAA	Federal Aviation Administration

Acronym	Term
FACI	First Article Configuration Inspection
FAD	Formulation Authorization Document
FAR	Federal Acquisition Regulation
FAR	Federal Acquisition Regulations
FAR	Flight Acceptance Review
FC	Fracture-Critical
FCDC	Flexible Confined Detonation Cord
FDC	Fire Department Connection
FDIR	Fault Detection Isolation and Recovery
FDO	Flight Dynamics Officer
FED-STD	Federal Standard
FEH	Facilities Engineering Handbook
FEMA	Federal Emergency Management Agency
FEP	Fluorinated Ethylene Propylene
FFRDC	Federally Funded Research and Development Center
FHA	Facility Hazard Analysis
FHA	Flight Hazard Area
FIFO	First-In-First-Out
FISMA	Federal Information Security Management Act
FM	Factory Mutual (Data Sheets)
FM	Factory Mutual
FM	Frequency Modulation
FMEA	Failure Modes and Effects Analysis
FMEA/CIL	Failure Modes and Effects Analysis/Critical Items List
FMECA	Failure Modes, Effects, and Criticality Analysis
FOC	Fiber Optic Cable
FOD	Foreign Object Debris
FOIA	Freedom of Information Act
FOTP	Fiber Optic Test Procedure
FPGA	Field Programmable Gate Array

Acronym	Term	Acronym	Term
FRACAS	Failure Reporting and Corrective Action System	GPMC	Governing Program Management Council
FRI	Facility Risk Indicator	Gr/Ep	Graphite Epoxy
FRL	Failure Rate Level	GSA	General Services Administration
FRR	Flight Readiness Review	GSE	Ground Servicing/Support Equipment
FS	Factor of Safety	GSE	Ground Support Equipment
FSI	Flame Spread Index	GSFC	Goddard Space Flight Center
FSMP	Facility Safety Management Plan	GSS	Ground Support System
FSO	Flight Safety Officer	GTO	Geosynchronous Transfer Orbit
FSS	Flight Safety System(s)	HA	Hazard Analysis
FTA	Fault Tree Analysis	HAD	Heat Actuated Device
FTS	Flight Termination System	HALT	Highly Accelerated Life Testing
GAO	General Accountability Office	HAN	Hydroxylammonium Nitrate
GAO	Government Accountability Office	HASC	Hazard Analysis Sub Committee
GAS	Get Away Special	HAST	Highly Accelerated Stress Testing
GB	Guidebook	HATI	Hazard Analysis Tracking Index
GEIA	Government Electronics & Information Technology Association	HAZMAT	Hazardous Materials
		HAZOP	Hazard and Operability Study
GEO	Geosynchronous Earth Orbit	HBM	Human Body Model
GFE	Government Furnished Equipment	HD	Hazard Division
GFE	Government-Furnished Equipment	HDBK	Handbook
GH$_2$	Gaseous Hydrogen	HDL	Hardware Description Language
GHe	Gaseous Helium	HE	High Explosive
GIDEP	Government Industry Data Exchange Program	HEA	Human-Error Analysis
GIDEP	Government-Industry Data Exchange Program	HEOMD	Human Exploration and Operations Mission Directorate
GMIP	Government Mandatory Inspection Point	HFD	Hazardous Fragment Distance
		HIF	Horizontal Integration Facility
GOES	Geostationary Operational Environmental Satellite	HIPAA	Health Insurance Portability and Accounting Act
GOP	Ground Operations Plan	HI-REL	High Reliability
GOTS	Government Off-The-Shelf	HLTR	Hazard List Tracking Record
GOTS	Government off-The-Shelf Software	HMS	Hazard Monitor System
GOV	Government Owned Vehicle	HMX	Cyclotetramethylenetetranitramine
GOX	Cryogenic Oxygen	HNS	Hexanitrostilbene

Acronym	Term	Acronym	Term
HOP	Hazardous Operating Procedure or Hazardous Operating Permit	IFSTA	International Fire Service Training Association
HOS	Hazardous Operations Support	IGSCC	Intergranular Stress Corrosion Cracking
HOWI	NASA Headquarters Office Work Instruction	ILD	Intraline Distance
HPM	High Performance Magazine	IMD	Intermagazine Distance
HQ	NASA Headquarters	INSRP	Interagency Nuclear Safety Review Panel
HQ	Headquarters	IPAO	Independent Program Assessment Office
HQR	Handling Qualities Rating		
HRCP	Human-Rating Certification Package	IPC®	Registered Trademark for IPC®-Association Connecting Electronic Industries
HRV	Hazard Resolution Verification		
HSPD-5	Homeland Security Presidential Directive	IPF	Integration Processing Facility
I	Importance Factor	IPR	Independent Peer Review
I&T	Integration and Test	IR	Insulation Resistance
IA	Investigating Authority	IRFNA	Inhibited Red Fuming Nitric Acid
IADC	Inter-Agency Space Debris Coordination Committee	IRT	Interim Response Team
		ISI	Inservice Inspection
IAEA	International Atomic Energy Agency	ISO	International Organization for Standardization
IAOP	Inter-Center Aircraft Operations Panel		
		ISS	International Space Station
IATA	International Air Transport Association	IST	Initial System Test
		IT	Information Technology
IBC	International Building Code	ITA	Independent Technical Authority
IBD	Inhabited Building Distance	ITAR	International Traffic in Arms Regulations
ICAO	International Civil Aviation Organization		
IDC	Insulation Displacement Connection	ITSDF	Industrial Truck Standards Development Foundation
IEC	International Electrotechnical Commission	IV&V	Independent Verification and Validation
IEEE	Institute of Electrical and Electronics Engineers	JAN	Joint Army Navy
		JAXA	Japan Aerospace Exploration Agency
IEEE	Institute of Electrical and Electronics Engineers, Inc.	JCGM	Joint Committee for Guides in Metrology
IESD	Internal Electrostatic Discharge	JIMO	Jupiter Icy Moons Orbiter
IFOSA	Institutional, Facility, Operational Safety Audit	JP	Jet Propellant
		JP-10	Hydrocarbon Turbine/Ramjet Fuel

Acronym	Term	Acronym	Term
JPL	Jet Propulsion Laboratory (a Federally Funded Research and Development Center)	LCC	Life-Cycle Cost
JPL	Jet Propulsion Laboratory, a Federally Funded Research Development Center	LDC	Lot Date Code
		LDE	Lifting Devices and Equipment
JPL	Jet Propulsion Laboratory (a Federally Funded and Research Development Center)	LDEC	Lifting Devices and Equipment Committee
		LDEM	Lifting Devices and Equipment Manager
JPL	NASA Jet Propulsion Laboratory	LED	Light Emitting Diode
JPL	Jet Propulsion Laboratory	LEL	Lower Explosive Limit
JPL (FFRDC)	Jet Propulsion Laboratory Federally-Funded Research Development Center	LEO	Low Earth Orbit
		LFCP	Lead-Free Control Plan
		LFU	Laser Firing Unit
JSC	Johnson Space Center	LH_2	Liquid Hydrogen
JSC	NASA Johnson Space Center	LHe	Liquid Helium
JSpOC	Joint Space Operations Center	LID	Laser Initiated Device
JTA	Joint Technical Architecture	LIFO	Last-In-First-Out
KDP	Key Decision Point	Li-Ion	Lithium Ion
KDP	Kennedy Documented Procedure	LIO	Laser Initiated ordnance
KHB	Kennedy Handbook	LIOS	Laser Initiated ordnance System
K_I	Stress Intensity	LLIS	Lessons Learned Information System
K_{Ic}	Plane-Strain Fracture toughness	LN_2	Liquid Nitrogen
K_{Ie}	Surface-Crack Tension Specimen Fracture toughness	LO_2	Liquid Oxygen
		LOC	Loss of Crew
K_{ISCC}	Stress-Corrosion Cracking Threshold	LOD	Letter of Delegation
KM	Knowledge Management	LOM	Loss of Mission
K_{max}	Maximum Stress Intensity Factor	LOX	Liquid Oxygen
KNPR	Kennedy NASA Procedural Requirements	LPS	Lightning Protection System
		LRF	Low-Risk Fracture
KSC	Kennedy Space Center	LRR	Launch Readiness Review
KTI	Kennedy Technical Instruction	LSIM	Launch Site Integration Manager
KVA	Kilovoltampere	LSO	Laser Safety officer
L	Liters	LSP	Launch Services Program
L/min	Liters per Minute	LSPM	Launch Services Program Manager
LBB	Leak Before Burst	LSRRR	Launch Safety Requirements Relief Request
LC	Life Cycle		

Acronym	Term
LWD	Liquid Withdrawal Device
M&P	Materials and Processes
MAC	Maximum Allowable Concentration
MADCAP	Multi-mission Automated Deepspace Conjunction Assessment Process
MAPTIS	Material and Process Technical Information System
MAWP	Maximum Allowable Working Pressure
MCE	Maximum Credible Event
MCR	Mission Concept Review
MDAA	Mission Directorate Associate Administrator
MDCP	Mechanical Damage Control Plan
MDMT	Minimum Design Metal Temperature
MDR	Mission Definition Review
MDT	Mean Downtime
MEM	Meteoroid Environment Model
MEO	Meteoroid Environment Office
MEO	Medium Earth Orbit
MEOP	Maximum Expected Operating Pressure
MER	Mars Exploration Rover
MFCO	Mission Flight Control Officer
MHE	Material Handling Equipment
MI	Mishap Investigator
MIB	Mishap Investigation Board
MIDEX	Medium Class Explorers
MIL-HDBK	Military Handbook
MIL-PRF	Military Performance
Mil-Spec	Military Specification
MIL-STD	Military Standard
MILVAN	Military Van
MISO	NSC Mishap Investigation Support Office
MIT	Mishap Investigation Team

Acronym	Term
MIT	Certified IPC® Master Trainer
MMH	Monomethylhydrazine
MMH	Maintenance Man-Hour
MMOD	Micrometeoroids and Orbital Debris
MOA	Memorandum of Agreement
MOP	Maximum Operating Pressure
MOTS	Modified Off-The-Shelf
MOTS	Modified off-The-Shelf Software
MOU	Memorandum of Understanding
MPCP	Mishap Preparedness and Contingency Plan
MPE	Maximum Predicted Environment; Maximum Permissible Exposure
MRB	Mission Readiness Briefing
MRB	Material Review Board
MRO	Mars Reconnaissance Orbiter
MSA	Mine Safety Appliances
MSDS	Material Safety Data Sheet
MSFC	Marshall Space Flight Center
MSFC	NASA Marshall Space Flight Center
MSO	Mission Support Offices
MSPSP	Missile System Prelaunch Safety Package
MTA	Maintenance Task Analysis
MTBF	Mean Time Between Failure
MTE	Measuring and Test Equipment
MTR	Manufacturer's Test Report
MTTF	Mean Time To Failure
MTTR	Mean Time To Repair
MUA	Materials Usage Agreement
MWAR	Mishap Warning-Action-Response
N_2H_4	Hydrazine
N_2O_4	Nitrogen Tetroxide
NACE	National Association of Corrosion Engineers

Acronym	Term
NAICS	North American Industrial Classification System
NANADARTS	NASA Advisory, Notice, and Alerts Distribution and Response Tracking System
NARS	NASA Advisory Reporting System
NAS	National Aerospace Standard
NAS	National Airspace System
NASA	National Aeronautics and Space Administration
NASA	National Aeronautic and Space Administration
NASA SP	NASA Special Publication
NASA TM	NASA Technical Memorandum
NASA/FLAGRO	NASA Fatigue Crack Growth Computer Program
NASA-HDBK	NASA Handbook
NASA-STD	NASA Standard
NASM	National Aerospace Standard Military
NBIC	National Board Inspection Code
NCSL	National Conference of Standards Laboratories
NDA	Nondisclosure Agreement
NDE	Nondestructive Examination
NDE	Nondestructive Evaluation
NDI	Nondestructive Inspection
NDT	Nondestructive Testing
NEC	National Electrical Code
NEC	National Electric Code
NEI	Non-Explosive Initiator
NEPA	National Environmental Policy Act
NEPP	NASA Electronic Parts and Packaging Program
NESC	NASA Engineering and Safety Center
NEW	Net Explosive Weight

Acronym	Term
NEWQD	Net Explosive Weight for Quantity Distance
NF	NASA Form
NFC	National Fire Code
NFPA	National Fire Protection Association
NFPA	National Fire Prevention Association
NFPA	National Fire Protection Act
NFS	NASA FAR Supplement
NFSAM	Nuclear Flight Safety Assurance Manager
NHB	NASA Handbook
NIMS	National Incident Management System
NIOSH	National Institute of Occupational Safety and Health
NIST	National Institute of Standards and Technology (formerly The National Bureau of Standards)
NIST	National Institute of Standards and Technology
NMI	NASA Management Instruction
NMIS	NASA Mishap Information System
NMO	NASA Management Office
NODIS	NASA Online Directives Information System
NODIS	NASA On-Line Directives Information System
NODIS	NASA On-Line Directives Information System
NOTAM	Notice to Airmen
NOTMAR	Notice to Mariners
NPD	NASA Policy Directive
NPG	NASA Procedures and Guidelines
NPR	NASA Procedural Requirements
NPR	NASA Procedure Requirements
NPS	Nominal Pipe Size
NPSL	NASA Parts Selection List

Acronym	Term	Acronym	Term
NPSL	NASA Parts Selection List	OES	Optical Emission Spectroscopy
NPT	National Pipe Thread	OIG	Office of Inspector General
NRC	Nuclear Regulatory Commission	OIIR	Office of International and Interagency Relations
NRRS	NASA Record Retention Schedules	OIS	Operational Information System
NRTL	Nationally Recognized Testing Laboratory	OIS	Operational Information System
NSC	NASA Safety Center	OMB	Office of Management and Budget
NSC	National Safety Council	OPLAN	Operations Plan
NSI	NASA Standard Initiator	OPR	Office of Primary Responsibility
NSRS	NASA Safety Reporting System	OPR	Office of Primary Responsibility
NSS	NASA Safety Standard	OPS	Office of Protective Services
NSS/GO	NASA Safety Standard/Ground Operations	ORR	Operational Readiness Review
NSTC	NASA Safety Training Center	OS&Y	Outside Screw and Yoke
NSTS	National Space Transportation System	OSC	Operations Safety Console
NTIS	National Technical Information Service	OSH	Occupational Safety and Health
NTS	NASA Technical Standard	OSHA	Occupational Safety and Health Administration
NTSB	National Transportation Safety Board	OSHA	Department of Labor, Occupational Health and Safety Administration
O&M	Operating and Maintenance	OSHA	Occupational Safety and Health Agency
O&SHA	Operating and Support Hazard Analysis	OSI/AMD	Office of Strategic Infrastructure, Aircraft Management Division
O&SHA	Operational and Support Hazard Analysis	OSMA	Office of Safety and Mission Assurance
OCE	Office of the Chief Engineer	OSP	Operations Safety Plan
OCF	Out of Controlled Flight	OSTP	Office of Science and Technology Policy
OCHMO	Office of the Chief Health and Medical Officer	OTDR	Optical Time Domain Reflectometry
OCM	Original Component Manufacturer	OTS	Off-The-Shelf
OCOM	Headquarters Office of Communications	OTV	Operation Television
OD	Outside Diameter	P/F/A	Problems/Failures/Anomalies
ODA	Orbital Debris Assessment	PA	Public Address
ODAR	Orbital Debris Assessment Report	PAD	Percussion-Activated Device
ODPO	NASA Orbital Debris Program Office	PAO	Center Public Affairs Office
OEM	Original Equipment Manufacturer	PAPL	Program or Project Approved Parts List
		PAR	Pre-flight Acceptance Review

Acronym	Term	Acronym	Term
Pb	Lead	POC	Point of Contact
PBAN	Polybutadiene Acrylonitrile	POV	Privately Owned Vehicle
P_C	Probability of Casualty	PPD	Partner Program Directive
PCB	Printed Circuit Board	PPE	Personal Protective Equipment
PCB	Parts Control Board	PPF	Payload Processing Facility
PCN	Product Change Notice	PPO	Planetary Protection Officer
PD	Presidential Directive	PQASP	Program/Project Quality Assurance Surveillance Plan
PD/NSC	Presidential Directive/National Security Council	PR	Problem Report
PDN	Product Discontinuance Notification	PRA	Probabilistic Risk Assessment
PDR	Preliminary Design Review	PRACA	Problem Reporting and Corrective Action System
PEM	Plastic Encapsulated Microcircuit		
PER	Preliminary Engineering Report	PRD	Pressure Relief Device (to include pressure relief valve, rupture disc, or other device)
PES	Potential Explosive Site		
PET	Polyethylene Terephthalate	PRE-SHIP	Review Prior to Shipment to Launch Site.
PETN	Pentaerythritoltetranitrate		
PFA	Plastic Films, Foams, and Adhesive Tapes	PRP	Personnel Reliability Program
PFA	Probability of False Acceptance	PRT	Platinum Resistance Thermometer
PHA	Preliminary Hazard Analysis	PRV	Pressure Relief Valve
PHE	Propellant Handlers Ensemble	PSA	Parts Stress Analysis
PHL	Preliminary Hazard List	PSI	Pounds Per Square Inch
Pi	Probability of Impact	PSI	Payload Safety Introduction
PIN	Part or Identifying Number	PSIB	Payload Safety Introduction Briefing
PIND	Particle Impact Noise Detection	PSM	Process Safety Management
PIV	Post Indicator Valve	PSM	Pressure Systems Manager
PL	Public Law	PSWG	Payload Safety Working Group
PLAR	Post-Launch Assessment Review	PTFE	Polytetrafluoroethylene
PLC	Programmable Logic Controller	PTH	Plated Through Hole
PLD	Programmable Logic Devices	PTR	Public Traffic Route
PM	Performance Measure	PTR	Program Trouble Report
PM	Project Manager	PTRD	Public Traffic Route Distance
PMC	Program Management Council	PV/S	Pressurized Vessels and Pressurized Systems
PMPCB	Parts, Materials, and Processes Control Board	PVDF	Polyvinylidene Fluoride

Acronym	Term	Acronym	Term
PVS	Pressure Vessels and Pressurized Systems	RHAP	Radiation Hardness Assurance Plan
PWA	Printed Wiring Assembly	RIDM	Risk-Informed Decision Making
QA	Quality Assurance	RLV	Reusable Launch Vehicle
QAAR	Quality Audit, Assessment, and Review	RMA	Reliability, Maintainability, and Availability
QASAR	Quality and Safety Achievement Recognition	RMP	Risk Management Plan
QASP	Quality Assurance Surveillance Plan	RMS	Root Mean Square
QD	Quantity Distance	ROSE	Resistivity of Solvent Extract
QML	Qualified Manufacturers List	RP	Rocket Propellant
QMS	Quality Management System	RPCP	Red Plague Control Plan
QPL	Qualified Product List	RPO	Radiation Protection officer
R & D	Research and Development	RSC	Radiation Safety Committee (Eastern Range)
R&M	Reliability and Maintainability	RSM	Range Safety Manual
RAC	Risk Assessment Code	RSO	Range Safety Officer
RAC	Risk Assessment Classification	RSO	Radiation Safety Officer
RADCC	Radiological Control Center	RT	Radiographic Testing
RADSAFCOM	Radiation Safety Committee (Western Range)	S&A	Safe and Arm Device
RBD	Reliability Block Diagram	S&A	Status and Alert
RBDA	Reliability Block Diagram Analysis	SA	Software Assurance
RCC	Range Commanders Council	SAARIS	Surveys, Audits, Assessments, Reviews Information System
RCM	Reliability-Centered Maintenance	SAE	SAE International
RDX	Cyclotrimethylenetrinitramine	SAIA	Scaffold and Access Industry Association
REDAA	Requirements Evaluation and Documentation Assessment & Analysis	SAS	Safety Analysis Summary
RF	Radio Frequency	SAS	Supplier Assessment System
RFI	Radio Frequency Interference	SATERN	System for Administration, Training, and Educational Resources for NASA
RFP	Request for Proposal	SBU	Sensitive But Unclassified
RFP	Request for Proposals	SCAPE	Self-Contained Atmospheric Protective Ensemble
RH	Relative Humidity	SCCB	Software Configuration Control Board
RHA	Radiation Hardness Assurance	SCCSF	Safety Critical Computer System Function
RHAE	Radiation Hardness Assurance Engineer	SCD	Source Control Drawing

Acronym	Term
SCM	Software Configuration Management
SCN	Specification Change Notice
SCS	Space Control Squadron
SDI	Smoke Development Index
SDP	Safety Data Package formerly Mspsp
SDR	System Definition Review
SDS	Safety Data Sheet
SEE	Single Event Effects
SEE	Single Event Effect
SEI	Software Engineering Institute
SEMP	Systems Engineering Management Plan
SER	Safety Evaluation Report
SEU	Single Event Upset
SFI	Software Formal Inspection
SFIS	Software Formal Inspection Standard
SFP	Single Failure Point
SGI	Service Group I
SHA	System Hazard Analysis
SI	Systeme Internationale, or Metric System of Measurement
SI	International System of Units
SIR	System Integration Review
SIR	System Interface Review
SLC	Space Launch Complex
SMA	Safety and Mission Assurance
SMA TA	Safety and Mission Assurance Technical Authority
SMARTS	Safety and Mission Assurance Requirements Tracking System
SME	Subject Matter Expert
SMEX	Small Explorer
SMO	Systems Management office
SMSR	Safety and Mission Success Review
SMT	Surface Mount Technology

Acronym	Term
Sn	Tin
SNT-TC	Society for Nondestructive Testing-Testing Certification
SOC	System On Chip
SoC	System-On-a-Chip
SOP	Standard Operating Procedure
SOP	Standard Operating Procedures
S-P	Severity-Probability
SPECSINTACT	Specifications Kept Intact (Guide Specification System)
SPF	Single Point Failure
SPR	Software Problem Report
SPS	Service Preparation Subsystem
SQA	Software Quality Assurance
SR	Safety Review
SR	Sounding Rocket
SR&QA	Safety, Reliability, and Quality Assurance
SRM&QA	Safety, Reliability, Maintainability, and Quality Assurance
SRR	Systems Requirements Review
SRR	System Requirements Review
SRS	Software Requirements Specification
SSA	Space Situational Awareness
SSCA	Software Safety Criticality Assessment
SSHA	Subsystem Hazard Analysis
SSHA	Sub System Hazard Analysis
SSN	Space Surveillance Network
SSO	Sun Synchronous Orbit
SSP	Space Shuttle Program
SSP	System Safety Plan
SSPP	System Safety Program Plan
SSTP	System Safety Technical Plan
START	NASA Standards and Technical Assistance Resource Tool

Acronym	Term
STD	Standard
STD	Software Test Description
STP	Standard Temperature and Pressure
STR	Software Trouble Reports
SW	Software
SW	Space Wing
SWI	Space Wing Instruction
T.O.	Technical Order
TA	Technical Authority
TA	Technical Authority
TAA	Technical Assistance Agreement
TBI	Through Bulkhead Initiator
THZ	toxic Hazard Zone
TIA	Telecommunications Industry Association
TID	Total Ionizing Dose
TIM	Technical Interchange Meeting
TLV	Threshold Limit Value
TM	Test Method
TML	Total Mass Loss
TMO	Transportation Management office
TNT	Trinitrotoluene
TO	Transistor Outline
TOPO	Trajectory Operations Officer
TP	Technical Paper
TRL	Technical Readiness Level
TRR	Test Readiness Review
TWA	Time Weighted Average
UAS	Unmanned Aircraft Systems
UAV	Unmanned Aerial Vehicle
UBC	Uniform Building Code
UDMH	Unsymmetrical Dimethylhydrazine
UDS	Universal Documentation System
UFAS	Uniform Federal Accessibility Standard

Acronym	Term
UFC	Uniform Fire Code
UL	Underwriters Laboratories
UMC	Uniform Mechanical Code
UN	United Nations
UNO	United Nations Organizations
US	United States
USAF	United States Air Force
USC	United States Code
USG	United States Government
USSTRATCOM	United States Strategic Command
UT	Ultrasonic Test
UV	Ultraviolet
V&V	Verification and Validation
VAFB	Vandenberg Air Force Base
VCS	Voluntary Consensus Standards
VCS	Voluntary Consensus Standard
Vdc	Volts, Direct Current
VIM	International Vocabulary of Metrology
VPP	Voluntary Protection Program
Vrms	Volts, Root Mean Square
WBS	Work Breakdown Structure
WBT	Web-Based Training
WCA	Worst Case Analysis
WFF	Wallops Flight Facility
W-NMTTC	Western NASA Manufacturing Technology Transfer Center
WOCC	Wing Operations Control Center
WR	Western Range
WRTB	Wire Rope Technical Board
WSTC	Workmanship Standards Technical Committee
WSTDA	Web Sling and Tie Down Association

4. DEFINITIONS

Term [Citing Document(s)]	Definitions
ΔV [STD 8719.14]	The change in the velocity vector caused by thrust measured in units of meters per second.
Abort [NPR 8705.2]	Same as Mission Abort. The forced early return of the crew to Earth when failures or the existence of uncontrolled catastrophic hazards prevent continuation of the mission profile and a return to Earth is required for crew survival. The crew is safely returned to Earth in the space system nominally used for entry and landing/touchdown.
Aboveground Magazine [STD 8719.12]	Any building or structure, except for an operating building, used for the storage of explosives. Aboveground magazines are all types of above grade (not earth-covered) magazines or storage pads. This includes storage in trucks, trailers, railcars, or cargo aircraft.
Accelerator [STD 8739.1]	A compounding material used in small amounts to increase the cure rate or to change the conditions of the reaction (e.g. cause it to occur at a lower temperature). Accelerators get consumed by the process.
Acceptable Risk [NPR 8715.3]	A level of risk, referred to a specific item, system or activity, that, when evaluated with consideration of its associated uncertainty, satisfies pre-established risk criteria.
Acceptance [STD 8709.20]	Agreement by the appropriate NASA Management Official to the change in the level of risk to programs, hardware and personnel and taking the responsibility for the potential outcome of any increase in risk.
acceptance tests [STD 8719.24 Annex]	the required formal tests conducted on hardware to ascertain that the materials, manufacturing processes, and workmanship meet specifications and that the hardware is acceptable for its intended use; also the formal required tests conducted on software to ascertain that the code meets specifications and is acceptable for its intended use.
Accessories, Connector [STD 8739.4]	Removable mechanical hardware, such as cable clamps, backshells, and screws, that are parts of the connector assembly in a harness. Removable mechanical hardware, such as cable clamps, backshells, and screws, that are parts of the connector assembly in a harness.
Accident [NPR 8715.3]	A severe perturbation to a mission or program, usually occurring in the form of a sequence of events, that can cause safety adverse consequences, in the form of death, injury, occupational illness, damage to or loss of equipment or property, or damage to the environment.
Accident Prevention [NPR 8715.3]	Methods and procedures used to eliminate the causes that could lead to a accident.
Accredited Laboratory [STD 6008]	A laboratory that has been recognized by the national and/or international standard-setting organizations to carry out specific tests competently according to established quality, management, administrative, and test method accreditation criteria.

Term [Citing Document(s)]	Definitions
Acquirer [STD 8719.13]	The entity or individual who specifies the requirements and accepts the resulting software products, including software embedded in a hardware system. The acquirer is usually NASA or an organization within the Agency but can also refer to the Prime contractor – subcontractor relationship as well.
Acquirer [STD 8739.9] [STD 8739.8]	The entity or individual who specifies the requirements and accepts the resulting software products. The acquirer is usually NASA or an organization within the Agency but can also refer to the Prime contractor – subcontractor relationship as well.
Acquirer [NPR 8000.4]	An Acquirer is a NASA organization that tasks another organization (either within NASA or external to NASA) to produce a system or deliver a service.
Adapter [STD 8739.4]	An intermediate device to provide for attaching special accessories or to provide special mounting means.
Adequate [STD 8719.11]	When referring to fire protection or life safety, the safeguards necessary to provide facilities and their occupants with protection against all known or recognized hazards.
Adhesive [STD 8739.5]	A polymeric compound, usually an epoxy, used to secure the optical fiber in a splice assembly or connector.
Adjudication [STD 8709.20]	The process that encompasses the process of review, concurrence, and approval of a request for relief from an Agency-wide SMA requirement. The process includes the approval or disapproval of the request by the Chief, Safety and Mission Assurance (or delegated approval authority) and acceptance or rejection of the change in risk and acceptance of the new risk level by the appropriate NASA management official. A request is adjudicated when all steps in the process are complete.
Advisor [NPR 8621.1]	Agency-level directive. A NASA directive with Agency-wide applicability; i.e., NASA Policy Directives (NPDs), NASA Procedural Requirements (NPRs), and NASA Interim Directives (NIDs). For a mishap investigation, an advisor is a Federal employee appointed to or engaged by the investigating authority in a non-voting role for domain knowledge and advice.
Aerozine 50 [STD 8719.24 Annex]	a 50-50 blend of hydrazine and unsymmetrical dimethylhydrazine.
Agency Coordinator [NPR 8735.1]	The NASA person that serves as the representative to the GIDEP in the Department of Defense to communicate with and provide the interfaces between the Agency and the GIDEP. This individual is a GIDEP Representative as defined by GIDEP.
Aggregate Risk [NPR 8000.4]	The cumulative risk associated with a given goal, objective, or performance measure, accounting for all significant risk contributors. For example, the total probability of loss of mission is an aggregate risk quantified as the probability of the union of all scenarios leading to loss of mission.

Term [Citing Document(s)]	Definitions
Aircraft Flight Mishap [NPR 8621.1]	a. A mishap occurrence associated with the operation of an aircraft that takes place between the time any person boards the aircraft with the intention of flight and all such persons have disembarked, and in which any person suffers a fatality or serious injury, or in which the aircraft receives substantial damage. b. A mishap occurrence associated with the operation of public or civil unmanned aircraft system that takes place between the time the system is activated with the purpose of flight and the time the system is deactivated at the conclusion of its mission, and in which any person suffers a fatality or serious injury, or in which the aircraft receives substantial damage.
Aircraft Ground Mishap [NPR 8621.1]	A mishap involving an aircraft or unmanned aircraft system that does not meet the threshold of an Aircraft Flight Mishap, and in which any person suffers a fatality or serious injury, or in which the aircraft receives substantial damage.
all-fire level [STD 8719.24 Annex]	the minimum direct current or radio frequency energy that causes initiation of an electroexplosive initiator or exploding bridgewire initiator or laser initiated device with a reliability of 0.999 at a confidence level of 95 percent as determined by a Bruceton test. Recommended operating level is all-fire current, as determined by test, at ambient temperature plus 150 percent of the minimum all-fire current.
allowable load (stress) [STD 8719.24 Annex]	the maximum load (stress) that can be allowed in a material for a given operating environment to prevent rupture or collapse or detrimental deformation; allowable load (stress) in these cases are ultimate load (stress), buckling load (stress), or yield load (stress), respectively.
Alternate Standards [STD 8739.6]	Workmanship requirements baseline that is offered by the supplier as a substitute for one or more of the workmanship standards referenced in Table 1 herein. Procedures are not alternate requirements standards.
Ammunition [STD 8719.12]	Projectiles, such as bullets and shot, together with their fuses and primers that can be fired from guns or otherwise propelled.
Analysis [STD 8739.9]	A method used to verify requirements that are more complex than can be verified by inspection, demonstration, or test. Analysis involves technical review of mathematical models, functional or operational simulation, equivalent algorithm tests, or other detailed engineering analysis.
Anomaly [STD 8729.1]	An unexpected event that is outside of certified design/performance specification limits or expectations.
antenna [STD 8719.24 Annex]	a device capable of radiating or receiving radio frequency energy.
Apogee [STD 8719.14]	The point in the orbit that is the farthest from the center of the Earth. The apogee altitude is the distance of the apogee point above the surface of the Earth.
Application [STD 8739.9]	A group of software elements: components or modules that share a common trait by which they are identified to the persons or departments responsible for their development, maintenance, or testing.
applied load [STD 8719.24 Annex]	the static or dynamic load applied to a structure, excluding load amplification factors.

Term [Citing Document(s)]	Definitions
applied load (stress) [STD 8719.24 Annex]	the actual load (stress) imposed on the structure in the service environment.
Appointing Official [NPR 8621.1]	The official authorized to appoint the investigating authority for a mishap or close call; accept the investigation of another authority; receive endorsements and comments from endorsing officials; and approve the mishap investigation report.
Approval [STD 8709.20]	Decision by the SMA TA that the request for relief is within NASA policy and may be implemented after the appropriate NASA Management official accepts the risk.
Approve [STD 8719.13]	The term approve or approval indicates that the responsible originating official, or designated decision authority, of a document, report, condition, waiver, deviation, etc. has agreed, via their signature, to the content and indicates the document is ready for release, baselining, distribution, etc. Usually, there will be one "approver" and several stakeholders who would need to "concur" for official acceptance of a document, report, waiver, etc. (for example, the project manager would approve the Software Development Plan, but SMA would concur on it.)
Approved Manufacturer [STD 6008]	A manufacturer that has passed an audit intended to verify that a company has the manufacturing capability and implemented quality management system with controlled processes that will ensure that products meet the requirements of applicable specifications.
Approved Mishap Investigation Report [NPR 8621.1]	The final mishap investigation report authorized for public release.
Approving authority or authorities [NPR 8715.7]	The organization(s) (internal and/or external to NASA) having the responsibility to grant approval/concurrence to perform processing and/or launch activities in their respective facilities, including acceptance of any associated risk.
Apsis [STD 8719.14]	The point in the orbit where a satellite is at the lowest altitude (perigee) or at the highest altitude (apogee). The line connecting apogee and perigee is the line of apsides.
Area Array Package [STD 8739.1]	A package with an X-Y grid interconnect pattern on the under-surface (i.e., ball grid array, column grid array, land grid array, pin grid array).
Argument Of Perigee [STD 8719.14]	The angle between the line extending from the center of the Earth to the ascending node of an orbit and the line extending from the center of the Earth to the perigee point in the orbit measured from the ascending node in the direction of motion of the satellite.
arm/disarm device [STD 8719.24 Annex]	an electrically or mechanically actuated switch that can make or break one or more ordnance firing circuits; operate in a manner similar to safe and arm devices except they do not physically interrupt the explosive train.
arming plug [STD 8719.24 Annex]	a removable device that provides electrical continuity when inserted in a firing circuit.
Ascending Node [STD 8719.14]	The point in the orbit where a satellite crosses the Earth's equatorial plane in passing from the southern hemisphere to the northern hemisphere.

Term [Citing Document(s)]	Definitions
Assessment [NPR 8715.3] [NPR 8715.7]	Review or audit process, using predetermined methods, that evaluates hardware, software, procedures, technical and programmatic documents, and the adequacy of their implementation.
Assessment [STD 8739.8]	An objective evaluation of performed processes or products and services against their applicable process descriptions, standards, procedures, and requirements.
Assessment [NPR 8705.6]	An activity that uses a set of concepts and principles, not a standard, to evaluate the accuracy, efficiency and/or effectiveness of an entity.
Assurance [NPR 8715.3]	Providing a measure of increased confidence that applicable requirements, processes, and standards are being fulfilled.
Attachment (for Industrial Trucks) [STD 8719.9]	Devices other than conventional forks or load backrest extensions, mounted permanently or temporarily on the elevating mechanism of an industrial truck for handling the load. Common types of attachments include but are not limited to fork extensions, clamps, rotating devices, side shifters, and booms.
Audit [NPR 8715.3]	Formal review to assess compliance with hardware or software requirements, specifications, baselines, safety standards, procedures, instructions, codes, and contractual and licensing requirements.
Audit [NPR 8715.7]	A formal review to assess compliance with hardware or software requirements, specifications, baselines, safety standards, procedures, instructions, codes, and contractual and licensing requirements.
Audit [STD 6008]	A systematic, independent, and documented process to verify that a company has the capability to manufacture fasteners with documented and controlled processes that meet the requirements of this standard.
Audit [STD 8739.8]	An examination of a work product or set of work products performed by a group independent from the developers to assess compliance with specifications, standards, contractual agreements, or other criteria. [Based on IEEE 610.12, IEEE Standard Glossary of Software Engineering Terminology]
Audit [NPR 8705.6]	A formal evaluation of compliance with SMA policies, procedures, processes, requirements, specifications, baselines, standards, instructions, codes, and contractual and licensing requirements.
Audit [STD 8739.9]	An examination of a work product or set of work products performed by a group independent from the developers to assess compliance with specifications, standards, contractual agreements, or other criteria.
Audit, Review, and Assessment Point of Contact [NPR 8705.6]	A person from the organization being audited, reviewed, or assessed who ensures that the audited organization is prepared for the audit, review, and assessment; coordinates the audit schedule with the audit, review, and assessment team; ensures that the appropriate personnel from the audited organization are available during the audit and can support the audit schedule; ensures that the audit, review, and assessment team has resources onsite to enable the completion of the audit (such as working space and information technology support).

Term [Citing Document(s)]	Definitions
Audit, Review, and Assessment Report [NPR 8705.6]	A document providing a record of an audit, review, or assessment results.
Audit, Review, and Assessment Team [NPR 8705.6]	A team comprising subject matter experts from NASA Headquarters, NASA Centers, and, if necessary, non-NASA organizations selected to conduct the SMA audits, reviews, and assessments.
Author [STD 8739.9]	The individual who created or maintains a work product that is being inspected. (see Wiegers 2008)
Authority Having Jurisdiction (AHJ) [STD 8719.11]	Refers to the individual(s) at the NASA Centers and Headquarters responsible for implementing the fire safety provisions of NPR 8715.3, "NASA General Safety Program Requirements," and with the authority for "approving/concurring in" associated installations, procedures, and equipment.
Automated [NPR 8705.2]	Automatic (as opposed to human) control of a system or operation.
Autonomous [NPR 8705.2]	Ability of a space system to perform operations independent from any Earth-based systems. This includes no communication with, or real-time support from, mission control or other Earth systems.
Autonomous Flight Safety System (AFSS) [STD 8719.25]	An onboard system that includes all hardware and software needed to make a flight termination decision (or other safety decision) and initiate actions that end vehicle flight (or otherwise restrict vehicle flight) without ground-based intervention. An Autonomous Flight Termination System (AFTS) is a type of AFSS.
Auxiliary Payload [NPR 8715.7]	A small satellite (e.g., CubeSats, Nanosatellites, Picosatellites) that does not interfere with the primary payload mission.
Availability [NPR 8715.3]	Measure of the percentage of time that an item could be used as intended.
Availability, Inherent (Ai) [STD 8729.1]	The percentage of time that a system or group of systems within a unit are operationally capable of performing an assigned mission with respect only to operating time and corrective maintenance time. It excludes logistics time, waiting or administrative downtime, and preventive maintenance downtime. It includes corrective maintenance downtime. Inherent availability is generally derived from analysis of an engineering design and is calculated as the Mean Time To Failure (MTTF) divided by the MTTF plus the Mean Time To Repair (MTTR). It is based on quantities under control of the designer.
Availability, Operational (Ao) [STD 8729.1]	The percentage of time that a system or group of systems within a unit are operationally capable of performing an assigned mission and can be expressed as uptime/(uptime+downtime). It includes logistics time, ready time, and waiting or administrative downtime, and both preventive and corrective maintenance downtime. This value is equal to the Mean Time Between Failure (MTBF) divided by the MTBF plus the Mean Downtime (MDT). This measure extends the definition of availability to elements controlled by the logisticians and mission planners such as quantity and proximity of spares to the hardware item. Ao is the quantitative link between readiness objectives and supportability.

Term [Citing Document(s)]	Definitions
Back-lit [STD 8739.5]	A method of illuminating the fiber end-face by launching incoherent light into the optical fiber core through the opposite end of the fiber.
Backscatter [STD 8739.5]	The return of a portion of scattered light to the input end of a fiber; the scattering of light in the direction opposite to its original propagation.
Backshell [STD 8739.4]	Are a connector accessory that are installed onto the rear connector accessory threads of plug or receptacle connectors to provide mechanical protection to the individual harness wires entering the back of the connector.
Barrel (Contact Wire Barrel) [STD 8739.4]	The section of contact that accommodates the stripped conductor.
Barricade [STD 8719.12]	An intervening approved barrier, natural or artificial, of such type, size, and construction as to limit, in a prescribed manner, the effect of an explosion on nearby buildings or exposures.
Barrier [NPR 8621.1]	A physical device intervention (e.g., a guardrail) or an administrative intervention that can provide procedural separation in time and space (e.g., lock-out/tag-out procedure) used to reduce risk of the undesired outcome.
Batch [STD 8739.1]	That quantity of material that was subjected to unit chemical processing or physical mixing, or both, designed to produce a product of substantially uniform characteristics.
battery capacity [STD 8719.24 Annex]	(1) rated capacity: the capacity assigned by the battery manufacturer based on a set of specific conditions such as discharge temperature, discharge current, end of discharge voltage, and state of charge at start of discharge; (2) measured capacity: the capacity determined by the specific qualification tests, including any time the battery is under load during qualification; the end of discharge voltage is the minimum voltage that flight termination system components have been qualified to.
Bay [STD 8719.12]	A location (examples: room, cubicle, cell, work area) that affords the level of safety and protection appropriate to the material and activity involved.
Below-the-Hook Lifting Device [STD 8719.9]	A device used for attaching a load to a hoist or other lifting mechanism. The device may consist of or contain components such as slings, hooks, and rigging hardware that are addressed by ASME B30 volumes or other standards. Common types include spreader bars, beam clamps, barrel lifters, and vacuum lifts. Some of these devices may be referred to as structural slings.
Bend Radius, Long Term [STD 8739.5]	The minimum radius to which a cable, without tensile load, can be bent for its lifetime without causing broken fibers, a localized weakening of the fibers, or a permanent increase in attenuation.
Bend Radius, Short Term [STD 8739.5]	The minimum radius to which a cable can be bent while under the maximum installation load without causing broken fibers, a localized weakening of the fibers, or a permanent increase in cable attenuation.

Term [Citing Document(s)]	Definitions
Birdcaging. [STD 8739.4]	The radial expansion of individual strands in a stranded conductor (bowing outward) that can occur in the exposed portion of the conductor between the insulation strip and termination point.
Blast Overpressure [STD 8719.12]	The pressure, exceeding the ambient pressure, manifested in the shock wave of an explosion.
Blister [STD 8739.1]	Undesirable rounded elevation of the surface of a polymer, whose boundaries may be more or less sharply defined.
Bonding [STD 8719.12]	The process of controlling static electric hazards by connectiong two or more conductive objects together by means of a conductor so they are at the same electrical potential, but not necessarily at the same potential as the earth.
Bonding [STD 8739.1]	Method for joining surfaces of parts or materials using a polymer.
Braid [STD 8739.4]	A fibrous or metallic group of filaments interwoven to form a protective covering over one or more wires.
Brake [STD 8719.9]	A device used for retarding or stopping motion.
Breakout [NPR 8705.2]	During proximity operations, the ability to maneuver one or more vehicles to a safe separation distance.
Breakout [STD 8739.4]	The separation of a conductor or group of conductors from the main body of wires in a harness.
brittle fracture [STD 8719.24 Annex]	(1) a type of failure mode in structural materials that usually occurs without prior plastic deformation and at extremely high speed, (2) a type of failure mode such that burst of the vessel is possible during cycling [normally this mode of failure is a concern when cycling to the maximum expected operating pressure (MEOP) or when the vessel is under sustained load at MEOP], and (3) a type of fracture that is characterized by a flat fracture surface with little or no shear lips (slant fracture surface) and at average stress levels below those of general yielding.
brittle materials [STD 8719.24 Annex]	see materials, brittle.
Bubble Pack [STD 8739.4]	A laminated plastic sheet that is formed with patterned air entrapment ("bubbles"). The bubbles provide excellent cushioning for anything enclosed between layers of the material.
Buddy System [NPR 8715.3]	An arrangement used when risk of injury is high, where personnel work in pairs, with one person in the pair stationed nearby, not directly exposed to the hazard, to serve as an observer to render assistance if needed.
Buffer [STD 8739.5]	A material applied over the coating that may be used to protect an optical fiber from physical damage, providing mechanical isolation or protection, or both.

Term [Citing Document(s)]	Definitions
burst factor [STD 8719.24 Annex]	a multiplying factor applied to the MEOP to obtain the design burst pressure; synonymous with ultimate pressure factor.
Cable [STD 8739.4]	A shielded single conductor or a combination of conductors insulated from one another (multiple conductor).
Cable Clamp [STD 8739.4]	A mechanical clamp attached to the wire entrance of a connector to support the cable or wire bundle, provide stress relief, and absorb vibration and shock.
Cable, Coaxial [STD 8739.4]	A cable in which an insulated conductor is centered inside another. The outer conductor is usually a metal braid or metal sheath. Braided cables usually have an outer insulating jacket over the braid. Coaxial cables are used primarily for transmission of RF signals.
Cable, Shielded [STD 8739.4]	One or more insulated conductors covered with a metallic outer covering, usually a metal braid.
Calibration System [STD 8739.12]	The set of interrelated or interacting elements necessary to maintain the measurement performance of measuring and test equipment to defined requirements. (ANSI/NCSL Z540.3)
Can [STD 8709.20]	Good practices, guidance, or options are specified with the nonemphatic verbs "should," "may," or "can" (from NASA-STD 0005).
Candidate Risk [NPR 8000.4]	A potential risk that has been identified and is pending adjudication by the affected programmatic or institutional authority.
Casualty [STD 8719.25]	An injury requiring overnight hospitalization or worse, including death. For the purpose of casualty modeling, any injury that, due to its severity, qualifies as a Level-3, 4, 5, or 6 injury per the Abbreviated Injury Scale (AIS), Association for the Advancement of Automotive Medicine, would be counted as a casualty.
Catalyst [STD 8739.1]	A substance that changes the rate of a chemical reaction without undergoing permanent change in its composition or getting consumed by the process.
Catastrophic [NPR 8715.3]	(1) A hazard that could result in a mishap causing fatal injury to personnel, and/or loss of one or more major elements of the flight vehicle or ground facility. (2) A condition that may cause death or permanently disabling injury, major system or facility destruction on the ground, or loss of crew, major systems, or vehicle during the mission.
Catastrophic [STD 8719.13]	[1] A condition causing fatal injury to personnel, and/or loss of one or more major elements of the flight vehicle or ground facility. A condition that may cause death or permanently disabling injury, major system or facility destruction on the ground, or loss of crew, major systems, or vehicle during the mission. [2] Loss of human life or permanent disability; loss of major system; loss of vehicle; loss of ground facility; severe environmental damage.
Catastrophic Event [NPR 8705.2]	An event resulting in the death or permanent disability of a crew member or passenger or an event resulting in the unplanned loss/destruction of a major element of the crewed space system during the mission that could potentially result in the death or permanent disability of a crew member or passenger.

Term [Citing Document(s)]	Definitions
Catastrophic Hazard [NPR 8705.2]	Any hazard that, when uncontrolled, results in a catastrophic event.
Catastrophic Hazard [STD 6008]	A hazard that can result in loss of life, a disabling injury, or the loss of spaceflight hardware (Space Shuttle, Space Station, Crew Launch Vehicle, Crew Exploration Vehicle, or Government-furnished Equipment), ground support equipment, ground facilities, or program-critical equipment.
Catastrophic hazard [NPR 8715.7]	A hazard, condition or event that could result in a mishap causing fatal injury to personnel and/or loss of spacecraft (payload), launch vehicle, or ground facility.
catastrophic hazard (payloads post-launch) [STD 8719.24 Annex]	a payload-related hazard, condition or event occurring post-launch (airborne) through payload separation that could result in a mishap causing fatal injury (including fatal injuries to the public) or loss of flight termination system.
catastrophic hazard (payloads prelaunch) [STD 8719.24 Annex]	a payload-related hazard, condition, or event occurring prior to launch (on ground) that could result in a mishap causing fatal injury to personnel or loss of spacecraft, launch vehicle, or ground facility.
catastrophic hazard (sample return to earth) [STD 8719.24 Annex]	a sample recovery-related hazard, condition or event occurring during sample recovery operations that could result in a mishap causing fatal injury to personnel.
Cause [NPR 8621.1]	An event or condition resulting in an effect. Anything that shapes or influences the outcome. A cause must precede and be necessary and sufficient on its own to bring about the undesired outcome of a mishap.
Cell (High Performance Magazine (HPM)) [STD 8719.12]	A reinforced concrete storage area in an HPM, separated from other cells by a specially designed non-propagation interior wall, with a removable reinforced concrete lid forming the roof. The entire HPM is earth-bermed.
Center Coordinator [NPR 8735.1]	A NASA person appointed by their organization to represent the Center and implement this NPR. This individual is a GIDEP Representative as defined by GIDEP.
Center Safety Office [NPR 8621.1]	The Center safety organization responsible for reporting and recording mishaps.
Center SMA [STD 8709.20]	The SMA office at a NASA Center/Facility. For component facilities, the SMA office of the parent NASA Center is the Center SMA office for purposes of the process defined in this document.
Center SMA Director [NPR 8705.6]	As used in this directive, this term includes all Center management personnel designated by the Center Director to implement SMA audits, reviews, and assessments requirements.
Certificate of Authorization (COA) or Waiver [NPR 8715.5]	A Certificate of Authorization (COA) or Waiver is a document issued by the FAA's Air Traffic Organization to a public operator (e.g., Government organizations, public universities, and law enforcement entities) for a specific activity for a specified period of time (i.e., temporary). The COA or Waiver will specify the operations that are permitted, define the area where the operations may be conducted, the period of time (i.e., temporary), and specify altitudes at which they may be conducted.

Term [Citing Document(s)]	Definitions
Certificate of Authorization or Waiver [STD 8719.25]	A Certificate of Authorization or Waiver is a document issued by the FAA's Air Traffic Organization to a public operator (e.g. Government organizations, public universities and law enforcement entities) for a specific unmanned aircraft activity for a specified period of time (i.e. temporary). The Certificate of Authorization or Waiver will specify the operations that are permitted, define the area where the operations may be conducted, and specify altitudes at which they may be conducted.
Certificate of Conformance (COC) [STD 6008]	A document that is signed by the fastener supplier to affirm that the product has met the requirements of the relevant specification(s), contractual requirements, and any other applicable regulations.
Certification [STD 8719.17]	The Center PSM's formal statement that a PVS complies with Agency requirements as specified in NPD 8710.5 and this standard, which require a documented process for assessment of integrity and risk, and compliance with applicable requirements.
Certification [STD 8739.4]	The act of verifying and documenting that personnel have completed required training, have demonstrated specified proficiency, and have met other specified requirements.
Certification Validation Test (CVT) [STD 6008]	Receiving inspection test(s) that are performed to assure conformance to the procurement specification requirements. For fasteners, this includes elemental analysis and mechanical property testing and inspection.
Certified Equipment [STD 8719.9]	Lifting device or equipment documented by the LDEM as complying with the design, construction, maintenance, test, and other requirements of this standard.
Chairperson [NPR 8621.1]	The individual in charge of a mishap investigation board or mishap investigation team.
Change House [STD 8719.12]	A building provided with facilities for employees to change to and from work clothes. Such buildings may be provided with sanitary facilities, drinking fountains, lockers, and eating facilities.
Checklist [STD 8739.9]	A list of procedures or items summarizing the activities required for an operator or technician in the performance of duties. A condensed guide. An on-the-job supplement to more detailed job instructions.
Chief, Safety and Mission Assurance Approval Required [STD 8709.20]	Approval of requests for relief that must come to the Chief, Safety and Mission Assurance.
Cladding [STD 8739.5]	The dielectric material surrounding the core of an optical fiber.
Classification of Hazard Contents [STD 8719.11]	Hazard contents of any building or structure are classified as low, ordinary, or high. • Low Hazard Contents: Such low combustibility that no self-propagating fire therein can occur. • Ordinary Hazard Contents: Likely to burn with moderate rapidity or to give off a large volume of smoke. • High Hazard Contents: Likely to burn with extreme rapidity or from which explosions are likely.

Term [Citing Document(s)]	Definitions
Classification Of Occupancies For Fire Suppression [STD 8719.11]	Occupancy classifications for this standard relate to sprinkler installations and their water supplies only. They are not intended to be a general classification of occupancy hazards. For purposes of determining required fire protection systems, occupancies will be protected according to their degree of hazard. Principal classifications, with typical examples, are listed under each category. (Note: The classification of unlisted occupancies will be based on an analysis of the hazards and a comparison with the definition and examples of listed occupancies). • Light Hazard Occupancies: Occupancies or portions of other occupancies where the quantity and combustibility of contents are low and fires with relatively low rates of heat release are expected. The facilities of NASA typically exceed this classification. • Ordinary Hazard Occupancies (Group 1): Occupancies or portions of other occupancies where combustibility of contents is low, quantity of combustibles is moderate, stock piles of combustibles do not exceed a height of 8 feet (2.44 meters), and fires with moderate rates of heat release are expected. Modest, scattered amounts of flammable liquids in closed containers are allowed in quantities up to 20 gallons (75.7 liters). The following are examples of Ordinary Hazard Occupancies (Group 1). • Auditoriums • Automobile parking garages • Cafeteria food preparation areas • Cafeteria seating areas • Classrooms • Clinics • Computer rooms • Drafting rooms and map making rooms • Electronic laboratories not normally using flammable liquids • File Rooms (files in metal cabinets) • Mechanical/electrical equipment room • Museums • Offices • Small storage rooms • Welding shops • Ordinary Hazard Occupancies (Group 2): Occupancies or portions of other occupancies where quantity and combustibility of contents are moderate, stockpiles do not exceed 12 feet (3.66 meters), and fires with a moderate rate of heat release are expected. Moderate, scattered amounts of flammable liquids in closed containers are allowable in quantities up to 50 gallons (189.3 liters). Small amount of flammable liquids may be exposed as required by normal operations. The following are examples of Ordinary Hazard Occupancies (Group 2). • Libraries • Mercantile • Magnetic tape libraries (tape in plastic cases and/or on plastic reels) • Model preparation areas • Piers and Wharves • Printing plants using inks having flash points at/or above 100 °F (37.9 °C) • Transformer vaults • Trash rooms • Vehicle repair garages • Warehouses (storage of noncombustible contents) • Woodworking shops • Extra Hazard Occupancies: Occupancies or portions of other occupancies where the quantity and combustibility of contents are very high or where flammable and combustible liquids, dust, lint, or other materials are present, introducing the probability of rapidly developing fires with high rates of heat release. The following are examples of Extra Hazard Occupancies: • Group 1: Aircraft hangars; Chemical laboratories; Engine test cells; Flammable and combustible liquids storage; Printing plants (using inks having flash points below 100 °F (37.9 °C); Upholstering with plastic foams; Warehouse with plastic foams; Warehouse (combustible contents stored not greater than 15 feet (4.57 meters) in piles of 12 feet (3.66 meters) in racks • Group 2: Flammable liquid spraying; Flow coating; Mobile home or modular building assemblies (where finished enclosure is present); Combustible interiors ; Open oil quenching; Plastics processing; Solvent cleaning; Paint dipping • Special Occupancies: Special Occupancies are facilities or areas which cannot be assigned a specific classification because of special protection requirements (refer to Chapter 10). This classification includes, but is not restricted to, the following occupancies. • High bay/payload processing areas • Launch facilities • Missile assembly areas • Ordnance storage/processing areas • Warehouses (high piled or high rack storage) • Combustible Liquid: A liquid having a flash point at or above 100 °F (37.9 °C).
Classification Yard [STD 8719.12]	A group of railroad tracks used for receiving, shipping, and switching railway cars.

Term [Citing Document(s)]	Definitions
Cleave [STD 8739.5]	The process of separating an optical fiber by a controlled fracture of the glass for the purpose of obtaining a fiber end that is flat, smooth, and perpendicular to the fiber axis.
Close Call [NPR 8621.1]	An event in which there is no or minor injury requiring first aid, or no or minor equipment or property damage (less than $20,000), but which possesses a potential to cause a mishap.
Closed-loop Reporting [NPR 8735.1]	Process by which a Program, Project, and Operational/Institutional Manager or their designee provides a usage assessment to a specific NASA Advisory or GIDEP Notice.
Coating [STD 8739.5]	A material put on a fiber during the drawing process to protect it from the environment.
Code PVS [STD 8719.17]	Pressure vessels and pressurized systems that are designed, fabricated, installed, code stamped, and maintained in strict conformance with the requirements of the VCS specified as applicable by the PSM.
Coefficient of Thermal Expansion (CTE) [STD 8739.1]	The measure of the fractional change in dimension per unit change in temperature.
Cognizant Safety Office [NPR 8621.1]	The responsible Safety Office within the host Center Safety and Mission Assurance Directorate that hosts the project or has been assigned safety and mission assurance accountability for the program.
Cold Flow [STD 8739.4]	Movement of insulation (e.g., Teflon) caused by pressure.
Collective Risk [STD 8719.25]	The total combined risk to all individuals exposed to one or more particular hazards during a specific period of time or event (a specific phase of flight). Unless otherwise noted, collective risk for a range flight operation is the mean number of casualties expected (Ec) during an established period or event (e.g., a launch) due to the combination of all hazards associated with the operation.
Collision Avoidance [STD 8719.25]	A process designed to prevent collisions between on-orbit tracked objects and launched or re-entering vehicles (including spent stages)/payloads by determining and implementing courses of action through careful analysis of validated conjunction assessments and satellite health and mission requirements. The process includes establishing wait periods in either the launch/entry window or spacecraft maneuvering based on validated conjunction assessments and accounts for uncertainties in spatial dispersions, arrival time of orbiting objects or the launch vehicle/payload, and modeling accuracy.
Combustible Material [STD 8719.12]	Any material which, when ignited, will sustain burning.
Commercial [STD 8739.10]	A classification for an assembly, part, or design for which the item manufacturer or vendor establishes performance, configuration and reliability, including design, materials, processes, and testing pursuant to market forces rather than by enforceable compliance to a government or industry standard.

Term [Citing Document(s)]	Definitions
Commercial and Government Entity (CAGE) Code [STD 8739.10]	An identifying code assigned by the Government that unambiguously identifies EEE part sources. A CAGE Code is required in order to conduct business with the Federal Government.
Commercial Launch [NPR 8715.5]	A service supplied by the private sector that provides the capability of placing a vehicle and any payload into a suborbital trajectory, Earth orbit, or into outer space.
Commercial Launch Service [STD 8719.25]	A service supplied by the private sector that provides the capability of placing a vehicle and any payload into a suborbital trajectory, Earth orbit, or into outer space.
Commercial Off-The-Shelf (COTS) [STD 8719.17]	Commercial items that require no unique Government modification or maintenance over the life cycle of the product to meet the needs of the procuring agency. A commercial item is one customarily used for nongovernmental purposes that has been or will be sold, leased, or licensed (or offered for sale, lease, or license) in quantity to the general public. An item that includes modifications customarily available in the commercial marketplace or minor modifications made to meet NASA requirements is still a commercial item. A custom engineered system, whether supplied by others or constructed by NASA, is not considered COTS.
Commingled [STD 6008]	A storage state where hardware (e.g., fasteners, inserts, etc.) from two or more different lots are co-located or stored in the same bin or other holding container.
Common Cause Failure [NPR 8705.2]	Failure of multiple items or systems due to a single event or common failure mode.
Common item [NPR 8735.1]	An item that has multiple applications versus a single or peculiar application.
Compatibility [STD 8719.12]	Chemical property of materials to coexist without adverse reaction for an acceptable period of time. Compatibility in storage exists when storing materials together does not increase the probability of an accident or, for a given quantity, the magnitude of the effects of such an accident. Storage compatibility groups are assigned to provide for segregated storage.
compatibility [STD 8719.24 Annex]	the ability of two or more materials or substances to come in contact without altering their structure or causing an unwanted reaction in terms such as permeability, flammability, ignition, combustion, functional or material degradation, contamination, toxicity, pressure, temperature, shock, oxidation, or corrosion.
Complete Traceability [STD 6008]	Documentation that demonstrates a solid chain of custody from the original fastener manufacturer through all intermediate distributors down to the buyer.
complex electronics [STD 8719.24 Annex]	encompasses programmable and designable complex integrated circuits. "Programmable" logic devices can be programmed by the user and range from simple chips to complex devices capable of being programmed on-the-fly. "Designable" logic devices are integrated circuits that can be designed but not programmed by the user.

Term [Citing Document(s)]	Definitions
Complex Item [NPR 8735.2]	A product that has quality characteristics not wholly visible in the end item, for which contract conformance cannot be determined through inspection, measurement, and/or test of the end item, and for which conformance can only be established progressively through the item's life by precise measurements, tests, and controls applied. Examples of complex items include assemblies, machinery, equipment, subsystems, systems, and platforms.
Complex Work [NPR 8735.2]	The design, manufacture, fabrication, assembly, testing, integration, operation, maintenance, refurbishment, or repair of complex items.
Component [STD 8719.12]	Any part of a complete item whether loaded with explosives (commonly called "live"), inert (not containing explosives), or empty.
Component [STD 8739.9]	A constituent element of a system or subsystem.
composite material [STD 8719.24 Annex]	the combinations of materials differing in composition or form on macro scale. The constituents retain their identities in the composite; normally, the constituents can be physically identified, and there is an interface between them.
Concept of Privilege [NPR 8621.1]	A level of confidentiality that a NASA (Federal employee) investigating authority or interim response team member may grant to a witness to an incident. Confidentiality means a witness is assured verbally and in writing that information provided during interviews or in a written statement will be protected by NASA to the extent provided by law.
Concur [STD 8719.13]	The term concur or concurrence means to agree and accept, via signature, the readiness and content of a condition, requirement, report, deviation, document, etc. This also implies that if the stakeholder (e.g. SMA) does not concur, that their sign-off is withheld and the document, waiver, deviation package, test report, hazard report, etc. is not to be considered acceptable until such changes are made to achieve agreement on the deliverable.
Concurrence [STD 8709.20]	Formal documentation of an agreement/recommendation/opinion, but with no authority to approve or accept risk.
Concurrent Operations [STD 8719.12]	Operations performed simultaneously and in close enough proximity that an incident with one operation could adversely influence the other.
Condition [NPR 8621.1]	A single as-found state.
Conductor [STD 8739.1] [STD 8739.4]	A lead or wire, solid, stranded, or printed wiring path serving as an electrical connection.
Configuration Control [STD 8739.9]	The systematic control of work products.
Configuration Item [STD 8739.8]	An aggregation of hardware, software, or both, that is established and baselined, with any modifications tracked and managed. [Based on IEEE 610.12, IEEE Standard Glossary of Software Engineering Terminology] Examples include requirements document, Use Case, or unit of code.

Term [Citing Document(s)]	Definitions
Configuration Item [STD 8739.9]	An aggregation of hardware, software, or both, that is established and baselined, with any modifications tracked and managed. Examples include requirements document, data block, Use Case, or unit of code.
Configuration Management [STD 8719.9] [STD 8719.17]	Process for establishing and maintaining consistency of a product's functional and physical characteristics, evaluating and authorizing any changes to those characteristics, and recording and documenting the characteristics and any changes to them to verify compliance with the product's configuration requirements throughout its life.
Configuration Management Organization (CMO) [STD 8709.20]	The collaborative configuration management effort shared between the Program/Project/Center and the Supplier (from NASA-STD 0005).
Conformal Coating [STD 8739.1]	A thin electrically nonconductive protective coating that conforms to the contours of the printed wiring assembly (PWA) or electronic assemblies.
Conformal Coating Test Specimen [STD 8739.1]	See Specimen, Conformal Coating Test.
Connector, Body [STD 8739.4]	The main portion of a connector to which contacts and other accessories are attached.
Connector, Grommet [STD 8739.4]	An elastomeric seal used on the cable side of a connector body to seal the connector against contamination and to provide stress relief.
Connector, Insert [STD 8739.4]	The part of a connector that holds the contacts in position and electrically insulates them from each other and the connector body.
Consequence [NPR 8000.4]	The key, possible negative outcome(s) of the current key circumstances, situations, etc., causing concern, doubt, anxiety, or uncertainty.
Consultant [NPR 8621.1]	For a NASA mishap investigation, a consultant is a non-Government subject matter expert engaged by the investigating authority for domain knowledge and analysis or opinion.
Contact [STD 8739.4]	The conductive element in a connector or other terminal device that mates with a corresponding element for the purpose of transferring electrical energy.
Contact Retention [STD 8739.4]	The axial load in either direction that a contact can withstand without being dislodged from its normal position within an insert or body.
Contact, Crimp [STD 8739.4]	A contact whose crimp barrel is a hollow cylinder that accepts the conductor. After a conductor has been inserted, a tool is used to crimp the contact metal firmly onto the conductor.
Contact, Insertable/Removable [STD 8739.4]	A contact that can be mechanically joined to or removed from an insert. Usually, special tools are used to insert (lock) the contact into place or to remove it.
Contact, Pin	Male-type contact designed to slip inside a socket contact.

Term [Citing Document(s)]	Definitions
[STD 8739.4]	
Contact, Socket [STD 8739.4]	A female-type contact designed to slip over a pin contact.
Contained Fastener [STD 6008]	A fastener that meets the criteria specified in NASA-STD-5019, Fracture Control Requirements for Spaceflight Hardware, paragraph 4.1.1.2.
Containment [STD 8719.25]	A range safety technique that precludes hazards from reaching the public, the workforce, or property that requires protection during normal and malfunctioning vehicle flight.
Contaminant [STD 8739.1]	An impurity or foreign substance present in a material that affects one or more properties of the material. A contaminant may be either ionic or nonionic. An ionic, or polar, compound forms free ions when dissolved in water, making the water a more conductive path. A nonionic substance does not form free ions, nor increase the water's conductivity. Ionic contaminants are usually processing residue such as flux activators, fingerprints, and etching or plating salts.
Contaminant [STD 8739.4]	An impurity or foreign substance present in a material that affects one or more properties of the material. A contaminant may be either ionic or nonionic. An ionic, or polar compound, forms free ions when dissolved in water, making the water a more conductive path. A nonionic substance does not form free ions, nor increase the water's conductivity. Ionic contaminants are usually processing residue such as flux activators, finger prints, and etching or plating salts.
contamination [STD 8719.24 Annex]	the introduction of impurities, undesirable material, suspect material, or material potentially out of specification that may render the system or equipment unusable for its intended purpose or in such a state that special measures need to be taken before the equipment or system can be restored to normal service.
Contingency [NPR 8621.1]	For planning, an emergency or urgent need that is regarded as unlikely but requiring some extent of pre-determined action if it occurred.
Contingency Management System (CMS) [NPR 8715.5] [STD 8719.25]	A system designed to manage the vehicle that provides a controlled response under the full set of circumstances defined by the mission's risk assessment. The system may be comprised of a set of elements within the vehicle, including but not limited to, manual control, autonomous control, and recovery capability.
Continuous Risk Management (CRM) [NPR 8000.4]	As discussed in paragraph 1.2.3, a systematic and iterative process that efficiently identifies, analyzes, plans, tracks, controls, and communicates and documents risks associated with implementation of designs, plans, and processes.

Term [Citing Document(s)]	Definitions
Contract [NPR 8735.2]	A mutually binding legal relationship obligating the seller to furnish the supplies or services (including construction) and the buyer to pay for them. It includes all types of commitments that obligate the Government to an expenditure of appropriated funds and that, except as otherwise authorized, are in writing. In addition to bilateral instruments, contracts include (but are not limited to) awards and notices of awards; job orders or task letters issued under basic ordering agreements; letter contracts; orders, such as purchase orders, under which the contract becomes effective by written acceptance or performance; and bilateral contract modifications. Contracts do not include grants and cooperative agreements covered by 31 U.S.C. 6301 et seq. or Space Act agreements covered by 41 U.S.C. 2473.
Contract [STD 8709.20]	Term is used interchangeably with agreement in this standard to indicate an agreement between a Supplier and a Program/Project/Center. This agreement could be between government organizations (e.g., task agreement) or between the Government and a business enterprise or academia (e.g., contract) (from NASA-STD 0005).
Contract [STD 8729.1]	An agreement between two or more parties, which is normally written and enforceable by law.
Contractor [STD 8729.1]	A party under contract to provide a product or service at a specified cost to another party (or parties) to the contract, also known as the customer(s).
Contributing Factor [NPR 8621.1]	An event or condition that may have contributed to the occurrence of an undesired outcome, but if eliminated or modified, would not on its own have prevented the occurrence.
Control [NPR 8621.1]	An active mechanism used to detect the initiating event or the hazard or both, and enable an active device (hardware, software, environmental, or human) to prevent or reduce the likelihood of the hazard affecting a target. Controls minimize effects of the initiating event by detecting and correcting them before bringing about an undesired outcome.
control authority [STD 8719.24 Annex]	a single commercial user on-site director and/or manager, a full time government tenant director and/or commander, or United States Air Force squadron/detachment commander responsible for the implementation of launch complex safety requirements.
Corrective Action [NPR 8621.1]	Any change that results in preventing, minimizing, or limiting the potential for occurrence of an incident (e.g., design processes, work instructions, workmanship practices, training, inspections, tests, procedures, specifications, drawings, tools, equipment, facilities, resources, material, and so on).
Corrective Action Plan Closure Statement [NPR 8621.1]	A final statement made by the appointing official documenting all corrective actions have been completed and the Corrective Action Plan is closed.
Corrective Actions [NPR 8735.1]	Changes to design processes, work instructions, workmanship practices, training, inspections, tests, procedures, software, specifications, drawings, tools, equipment, facilities, resources, or material that result in preventing, minimizing, or limiting the potential for recurrence.

Term [Citing Document(s)]	Definitions
Cost-Benefit Analysis [STD 8719.11]	A procedure in which the present value of future expenditures associated with the installation and maintenance of a fire safety system or device is related to the economic benefits of the facility or portion thereof that it is designed to protect. The technique is intended to determine the practicality of the installation of fire protection systems and must be limited to those situations where the possibility of loss of human life is low.
countdown [STD 8719.24 Annex]	the timed sequence of events that must take place to initiate flight of a launch vehicle.
Counterfeit [NPR 8735.1]	An unlawful or unauthorized reproduction, substitution, or alteration that has been knowingly mismarked, misidentified, or otherwise misrepresented to be an authentic, unmodified item from the original manufacturer, or a source with the express written authority of the original manufacturer or current design activity, including an authorized aftermarket manufacturer. Unlawful or unauthorized substitution includes used items represented as new, or the false identification of grade, serial number, lot number, date code, or performance characteristics.
Counterpoise [STD 8719.12]	A type of earth electrode system consisting of conductor cables buried around the structure to be protected. Generally, a counterpoise will have more surface area contacting the earth than ground rod systems.
Coupling Loss [STD 8739.5]	The optical power loss suffered when light is coupled from one optical device to another.
Crane [STD 8719.9]	A machine for lifting and lowering a load and moving it horizontally, with the hoisting mechanism an integral part of the machine.
Cratering Flux [STD 8719.14]	The number of impacts per square meter per year of objects which will leave a crater at least as large as a specified diameter.
Crew [NPR 8705.2]	Any human on board the space system during the mission that has been trained to monitor, operate, and control parts of, or the whole space system; same as flight crew.
Crew/Passenger Escape [NPR 8705.2]	See definition for escape.
Crew/Passenger Survival [NPR 8705.2]	Capability and ability to preclude crew/passenger fatality or permanent disability. The ability to keep the crew/passengers alive using such capabilities as abort, escape, safe haven, emergency egress, rescue and emergency medical, in response to an imminent catastrophic condition.
Crewed Element (of the Space System) [NPR 8705.2]	All system elements that are occupied by the crew/passengers during the space mission and provide life support functions for the crew/passengers. The crewed element includes all the subsystems that provide life support functions for the crew/passengers.
Crewed Space System [NPR 8705.2]	The crewed space system consists of all the system elements that are occupied by the crew/passengers during the space mission and provide life support functions for the crew/passengers (i.e., the crewed elements). The crewed space system also includes all elements physically attached to the crewed element during the mission. The crewed space system is part of the larger space system used to conduct the mission.

Term [Citing Document(s)]	Definitions
	· The following examples are provided for clarification of the definition of crewed space system as it relates to the Human-Rating Certification: · Application example 1: A launch vehicle for a crewed spacecraft on a NASA mission is part of the crewed space system for Earth ascent. In this example, the Human-Rating Certification applies to the launch vehicle and the spacecraft operating together as a crewed space system during the ascent phase of the reference mission. · Application example 2: A propulsion module, which is launched into space (un-crewed) and subsequently attached to a crewed spacecraft on a NASA mission, is part of the crewed space system for the Human-Rating Certification. As part of the certification, some of the requirements in this NPR will apply to the propulsion module during proximity operations with the crewed spacecraft. · Application example 3: The launch vehicle for the propulsion module in example 2 (when launched separately from crew) is not part of the crewed space system and will not be part of the Human-Rating Certification. · Application example 4: When the crew ingresses a vehicle for a launch attempt, the vehicle is physically connected to the launch pad. The entire launch pad is not considered part of the crewed system, but the specific launch pad systems that interact with the crewed vehicle are part of the crewed space system.
Crimp [STD 8739.4]	The physical compression (deformation) of a contact barrel around a conductor to make an electrical and mechanical connection to the conductor.
Crimping [STD 8739.4]	A method of mechanically compressing or securing a terminal, splice, or contact to a conductor.
Critical [NPR 8715.3]	A condition that may cause severe injury or occupational illness, or major property damage to facilities, systems, or flight hardware.
Critical [NPR 8735.2]	The condition where failure to comply with prescribed contract requirements can potentially result in loss of life, serious personal injury, loss of mission, or loss of a significant mission resource. Common uses of the term include critical work, critical processes, critical attributes, and critical items.
Critical [STD 8719.13]	[1] The condition where failure to comply with prescribed contract requirements can potentially result in loss of life, serious personal injury, loss of mission, or loss of a significant mission resource. Common uses of the term include critical work, critical processes, critical attributes, and critical items. [2] A condition that may cause severe injury or occupational illness, or major property damage to facilities, systems, or flight hardware.
Critical [STD 8739.10]	The condition where failure to comply with prescribed requirements can potentially result in loss of life, serious personal injury, loss of mission, or loss of a significant mission resource.
Critical (sub)System [NPR 8705.2]	A (sub)system is assessed as critical if loss of overall (sub)system function, or improper performance of a (sub)system function, could result in a catastrophic event or abort.

Term [Citing Document(s)]	Definitions
Critical Action [NPR 8705.2]	A critical action is defined as any operator action that, if performed in error during operations with zero or one system failures, would result in a catastrophic event or an abort.
critical condition [STD 8719.24 Annex]	the most severe environmental condition in terms of loads, pressures, and temperatures, or combination thereof imposed on structures, systems, subsystems, and components during service life.
critical facility/structure [STD 8719.24 Annex]	a hazardous facility or structure; a facility or structure used to store or process explosives, fuels, or other hazardous materials; a facility or structure used to process high value hardware; a facility or structure that contains or is used to process systems determined by Range Safety to be hazardous or critical; or a facility or structure determined by Range Safety to be critical.
Critical Functions [NPR 8705.2]	Mission capabilities or system functions that, if lost, would result in a catastrophic event or an abort.
critical hardware [STD 8719.24 Annex]	any hazardous or safety critical equipment or system; non-hazardous DoD high value items such as spacecraft, missiles, or any unique item identified by DoD as critical; non-hazardous, high value hardware owned by Range Users other than the DoD may be identified as critical or non-critical by the Range User; see also safety critical.
Critical hazard [NPR 8715.7] [STD 8719.24 Annex]	A hazard, condition or event that may cause severe injury or occupational illness, or major property damage to facilities, systems, or flight hardware.
Critical LDE [STD 8719.9]	Lifting Devices and Equipment used to perform Critical Lifts.
Critical Lift [STD 8719.9]	Lifts during which failure/loss of control presents an elevated risk of serious injury, loss of life, or loss of one-of-a-kind articles, high dollar items or major facility components whose loss would have serious programmatic or institutional impact. Lifts of high-value flight hardware and/or non-routine lifts (e.g., lift point below center of gravity) are usually classified as critical lifts, while lifts of small, improvised mini-satellites, for example, most likely would not be. Lifting and movement of flight hardware components packaged per applicable shipment specifications are typically not classified as critical lifts.
critical load [STD 8719.24 Annex]	a load consisting of critical hardware and/or any personnel.
Critical Nonconformance [NPR 8735.1]	A nonconformance that is likely to result in hazardous or unsafe conditions for individuals using, maintaining, or depending upon the supplies or services; or is likely to prevent performance of a vital agency mission.

Term [Citing Document(s)]	Definitions
Critical Operations Personnel [STD 8719.25]	Critical Operations Personnel include persons not essential to the specific operation (launch, entry, flight) currently being conducted, but who are required to perform safety, security, or other critical tasks at the launch, landing, or flight facility. Critical Operations Personnel are notified of the hazardous operation and either trained in mitigation techniques or accompanied by a properly trained escort. Critical Operations Personnel do not include individuals in training for any job or individuals performing routine activities such as administrative, maintenance, or janitorial. Critical Operations Personnel may occupy safety clearance zones and hazardous areas and need not be evacuated with the public. Critical Operations Personnel are included in the same risk category as Mission Essential Personnel.
Critical Single Failure Point [NPR 8715.3]	A single item or element, essential to the safe functioning of a system or subsystem, whose failure in a life or mission essential application would cause serious program or mission delays or be hazardous to personnel.
Critical Software [NPR 8705.2]	Any software component whose behavior or performance could lead to a catastrophic event or abort. This includes the flight software as well as ground-control software.
Critical Software Command [NPR 8715.3]	A command that either removes a safety inhibit or creates a hazardous condition.
Criticality (of a failure) [STD 8729.1]	A measure of the significance or severity of a failure on mission performance, hazards to material or personnel, and maintenance cost. Programs/projects typically establish their own criticality definitions and classifications.
Cross-cutting Risk [NPR 8000.4]	A risk that is generally applicable to multiple mission execution efforts, with attributes and impacts found in multiple levels of the organization or in multiple organizations within the same level.
cryogen [STD 8719.24 Annex]	a super cold liquid such as liquid nitrogen or oxygen.
Cure [STD 8739.1]	A chemical reaction that hardens and changes the physical properties of a material(s).
Cybersecurity Risk [NPR 8000.4]	Threats to and vulnerabilities of information or information systems and any related consequences caused by or resulting from unauthorized access, use, disclosure, degradation, disruption, modification, or destruction of information or information systems, including such related consequences caused by an act of terrorism. (From National Cybersecurity Protection Act of 2014.)
Damage [NPR 8621.1]	Either material or mission objective loss that is calculable as a Direct Cost (see Direct Cost of Mishap or Close Call).
Danger Area Information Plan [STD 8719.24 Annex]	an Eastern Range document prepared by Operations Safety specifying roadblocks and the fallback area associated with hazardous areas for each launch complex during launch operations.
Debris Flux [STD 8719.14]	The number of impacts per square meter per year expected on a randomly oriented planar surface of an orbiting space structure.

Term [Citing Document(s)]	Definitions
Debris Flux To Limiting Size [STD 8719.14]	The number of impacts per square meter per year of debris objects of a specified diameter or larger.
Debris Hazard [STD 8719.12]	A hazard resulting from any solid particle thrown by an explosion or other strong energetic reaction. For aboveground explosions, debris refers to secondary fragments.
decibel [STD 8719.24 Annex]	a unit of relative power; the decibel ratio between power levels, P1 and P2, is defined by the relation dB = 10 log (P1/P2).
dedicated [STD 8719.24 Annex]	serving a single function, such as a power source serving a single load.
Defect [STD 8739.9]	Any occurrence in a software product that is determined to be incomplete or incorrect relative to the software requirements, expectations for the software, and/or program standards.
Defect Classification [STD 8739.9]	The process where all defects identified during an inspection are classified by severity and type.
Deflagration [STD 8719.12]	A rapid chemical reaction in which the output of heat is sufficient to enable the reaction to proceed and be accelerated without input of heat from another source; a surface phenomenon with the reaction proceeding towards the unreacted material along the surface at subsonic velocity.
Degas [STD 8739.5]	The removal of entrapped bubbles from a viscous fluid by placing that fluid in a centrifuge or vacuum.
Delegated Agency [NPR 8735.2]	An organization providing Contract Administration Services (CAS) quality assurance support to NASA on designated contracts. Delegated agencies that provide NASA CAS support include the Defense Contract Management Agency (DCMA) and the Office of Naval Research (ONR).
Delegated agent [STD 8739.6]	NASA support contractor or alternate Federal Agency (e.g., Defense Contract Management Agency) that is formally delegated responsibility to perform Government Contract Quality Assurance functions in accordance with a written contract, task order, or Letter of Delegation.
Deliberation [NPR 8000.4]	In the context of this NPR, the formal or informal process for communication and collective consideration, by stakeholders designated in the Risk Management Plan, of all pertinent information, especially risk information, in order to support the decision maker.
Demolition/Demilitarization (Demil) [STD 8719.12]	Disarm, burn, explode, neutralize or any other action that will render the explosive/explosive device free of hazardous materials.
Dependability [STD 8729.1]	The ability to avoid service failures that are more frequent and more severe than is acceptable.

Term [Citing Document(s)]	Definitions
Derating [STD 8739.10]	Derating of a part is the intentional reduction of its electrical, mechanical and thermal stresses for the purpose of providing a margin between the applied stress and the actual demonstrated limit of the part capabilities.
Derived Requirement [STD 8709.20]	A requirement that is not a directed requirement.
Derrick [STD 8719.9]	An apparatus consisting of a mast or equivalent member held at the end by guys or braces, with or without a boom, for use with a hoisting mechanism and operating ropes.
Desiccant [STD 8739.10]	A hygroscopic substance that induces or sustains a state of dryness (desiccation) in its vicinity. A drying agent.
design burst pressure [STD 8719.24 Annex]	the calculated pressure (the analytical value that was calculated using an acceptable industry and/or government practice to determine its design pressure) that a component must withstand without rupture and/or burst to demonstrate its design adequacy in a qualification test; during qualification testing, the actual burst pressure for a tested component must demonstrate that the design burst pressure is less than the actual burst pressure; safety factors are based on design burst pressure, not actual burst pressure of a particular component.
Design Factor [STD 8719.9]	A numeric usually expressed as a ratio of the ultimate strength or yield strength to the rated capacity. It is used in calculations to account for variations found in the properties of materials, manufacturing tolerances, operating conditions, and design assumptions.
design load [STD 8719.24 Annex]	the value used by the manufacturer as the maximum load around which the device or equipment is designed and built based on specified design factors and limits. This is also the load referred to as the "Manufacturer's Rated Load." see also applied load.
design safety factor [STD 8719.24 Annex]	a factor used to account for uncertainties in material properties and analysis procedures; often called design factor of safety or simply safety factor.
Designated Person [STD 8719.9]	A person who is qualified and who has been selected or assigned (in writing) by the responsible organization to perform specific duties.
destabilizing pressure [STD 8719.24 Annex]	a pressure that produces comprehensive stresses in a pressurized structure or pressure component.
Destructive Physical Analysis (DPA) [STD 8739.10]	A series of inspections and tests performed on samples of an EEE part and resulting in damage to the samples. Usually part of a failure analysis or quality conformance inspection.
Deterioration [STD 8739.1]	(as in the context of the condition of stored polymer materials) A change in the material that can be observed prior to its use, or during use, that indicates it no longer meets its performance requirements. Deteriorated in this context includes degraded or separated.
detonating cord [STD 8719.24 Annex]	a flexible fabric tube containing a filler of high explosive material intended to be initiated by an electroexplosive device; often used in destruct and separation functions.

Term [Citing Document(s)]	Definitions
Detonation [STD 8719.12]	A violent chemical reaction within a chemical compound or mechanical mixture evolving heat and pressure that proceeds through the reacted material toward the unreacted material at a supersonic velocity.
detonation [STD 8719.24 Annex]	a violent chemical reaction within a chemical compound or mechanical mixture evolving heat and pressure that proceeds through the reacted material toward the unreacted material at a supersonic velocity; the result of the chemical reaction is exertion of extremely high pressure on the surrounding medium forming a propagating shock wave which is originally of supersonic velocity; a detonation, when the material is located on or near the surface of the ground, is normally characterized by a crater.
detonator [STD 8719.24 Annex]	an explosive device (usually an electroexplosive device) that is the first device in an explosive train and is designed to transform an input (usually electrical) into an explosive reaction.
detrimental deformation [STD 8719.24 Annex]	includes all structural deformations, deflections, or displacements that prevent any portion of the structure from performing its intended function or that reduces the probability of successful completion of the mission.
Developer [STD 6008]	Contractors who are not prime contractors and who design or build flight hardware. Examples include NASA-agreement entity organizations, colleges, schools, and universities.
development test [STD 8719.24 Annex]	a test to provide design information that may be used to check the validity of analytic technique and assumed design parameters, to uncover unexpected system response characteristics, to evaluate design changes, to determine interface compatibility, to prove qualification and acceptance procedures and techniques, or to establish accept and reject criteria.
Deviation [NPR 8715.3]	An authorization for temporary relief in advance from a specific requirement, requested during the formulation/planning/design stages of a program/project operation to address expected situations. OSHA refers to this as an alternate or supplemental standard.
Deviation [STD 8709.20]	A documented authorization releasing a program or project from meeting a requirement before the requirement is put under configuration control at the level the requirement will be implemented (from NASA Memo 7120-81, Appendix A, which updated NPR 7120.5D).
Deviation [STD 8739.10]	A specific written authorization, granted prior to the manufacture of a Configuration Item (CI), to depart from a particular requirement of a CI's current approved configuration for a specific number of units or a specified period of time.
Deviation [NPR 8705.2]	A documented authorization releasing a program or project from meeting a requirement before the requirement is put under configuration control at the level the requirement will be implemented. [NPD 7120.4 and NPR 7120.5]
Dielectric [STD 8739.1]	A material with a high resistance to the flow of electrical current, and which is capable of being polarized by an electrical field.
Dielectric Breakdown [STD 8719.12]	The failure of the insulating property of a material when the dielectric strength of the material has been exceeded and current flows through the material.

Term [Citing Document(s)]	Definitions
Diluent [STD 8739.1]	Any material that reduces the concentration of the fundamental resin; usually a liquid added to the resin to afford lower viscosity.
Direct Cost of Mishaps or Close Calls [NPR 8621.1]	For mishap classification, the sum of the costs (the greater value of actual or fair market value) of damaged property and/or destroyed property (public or NASA), or mission failure, actual cost of repair or replacement, labor (actual value of replacement or repair hours for internal and external or contracted labor), cost of the lost commodity (e.g., cost of the fluid lost from a ruptured pressure vessel), as well as resultant costs such as environmental decontamination, property cleanup, and restoration, or the estimate of these costs.
Directed Requirement [STD 8709.20]	An SMA requirement that has been imposed on NASA SMA as a flowdown of a requirement from a level higher to, or outside of OSMA. This includes requirements in standards which are called out as mandatory in the source of other directed requirements.
Direct-lit [STD 8739.5]	A method of illuminating the fiber end-face by projecting a light source onto the fiber.
Discrepancy [STD 8739.9]	A formally documented deviation of an actual result from its expected result.
Discrepancy Report [STD 8739.9]	An instrument used to record, research, and track resolution of a defect found in a baseline.
Displacement Damage Dose (DDD) [STD 8739.10]	Dose of radiation capable of causing displacement damage. Refers to the cumulative degradation resulting from the displacement of nuclei from their lattice position in a material due to ionizing or non-ionizing radiation.
Disposal [STD 8719.14]	An end-of-mission process for moving a spacecraft (if necessary) to an orbit considered acceptable for orbital debris limitation.
Dispositions (Risk) [NPR 8000.4]	(a) Accept: The formal process of justifying and documenting a decision not to mitigate a given risk. (See also Risk Acceptability Criterion). Note: A decision to "accept" a risk is a decision to proceed without further mitigation of that risk (i.e., despite exposure to that risk). (b) Close: The determination that a risk no longer exists (e.g., the underlying condition no longer exists), has become a problem and is now tracked as such, because the associated scenario likelihoods are low (e.g., the likelihood has been reduced below a defined threshold), or the associated consequences are low (e.g., the consequence has been reduced below a defined threshold). Note: Closing a risk due to low likelihood is still a risk acceptance decision. From a risk acceptance perspective, it is still necessary to account for the cumulative effects of risks closed due to low likelihood (see section l.). (c) Elevate: The process of transferring the decision for the management of an identified source of risk to the risk management structure at a higher organizational level. Note: Some organizational units within NASA use the term "escalate" to mean "elevate." (d) Mitigate: The modification of a process, system, or activity in order to reduce a risk by reducing its probability, consequence severity, or uncertainty, or by shifting its timeframe.

Term [Citing Document(s)]	Definitions
	Note: After mitigation, there will still be a need to accept any remaining risk and account for its contribution to the aggregate risk.
	(e) Research: The investigation of a risk in order to acquire sufficient information to support another disposition, i.e., close, watch, mitigate, accept, or elevate.
	(f) Watch: The monitoring of a risk for early warning of a significant change in its probability, consequences, uncertainty, or timeframe.
Distant Focusing Overpressure (DFO) [STD 8719.25]	An atmospheric phenomenon that can produce greatly enhanced overpressure due to sonic velocity gradients with respect to altitude. These enhanced overpressures can break windows in distant communities, which may result in personal injury. Distant focusing overpressure, sometimes referred to as far field blast overpressure, is of concern in the event of a large explosion on or around the launch pad and occurs only under certain meteorological conditions.
Distributor [STD 6008]	An enterprise that stocks the products of various manufacturers for resale and does not engage in manufacturing activity.
Dividing Wall [STD 8719.12]	A wall designed to prevent, control or delay propagation of a reaction involving explosives on opposite sides of the wall.
DOT Service [STD 8719.17]	Those uses of PVS covered by the regulations contained in 49 CFR 100 – 185, Pipeline and Hazardous Materials Safety Administration.
downrange [STD 8719.24 Annex]	the distance measured along a line whose direction is parallel to the projection of a launch vehicle's planned nominal velocity vector azimuth into a horizontal plane tangent to the ellipsoidal earth model at the launch vehicle sub-vehicle point; may also be used to indicate direction.
Drain Wire [STD 8739.4]	A wire that runs linearly along a foil shield wire or cable and is used to make contact with the shield. Grounding of foil shields is done with drain wires.
Dry Run [STD 8719.12]	Rehearsal of a process without the presence of the associated hazard. The level of dry run activities is dependent upon effect of change to the hazard level of process.
ductile failure [STD 8719.24 Annex]	see failure, ductile.
ductile fracture [STD 8719.24 Annex]	a type of failure mode in structural materials generally preceded by large amounts of plastic deformation and in which the fracture surface is inclined to the direction of the applied stress.
ductile materials [STD 8719.24 Annex]	see materials, ductile.
ductility [STD 8719.24 Annex]	the ability of a material to be plastically deformed without fracturing in tension or compression, respectively; two commonly used indices of ductility are the ultimate elongation and the reduction of cross-sectional area; the usual dividing line between ductility and brittleness is 5 percent elongation (See Metallurgy for Engineers, Mechanics of Materials, and Mechanical Engineering and Design in References.).

Term [Citing Document(s)]	Definitions
dudding [STD 8719.24 Annex]	the process of permanently degrading an electroexplosive initiator to a state where it cannot perform its designed function.
Dummy Load [STD 8719.9]	A test load used to simulate the real load; typically a test weight.
Dummy Rated Load [STD 8719.9]	A test load equal to the rated load of the device; typically a test weight.
Dunnage [STD 8719.12]	Inert (though possibly flammable) material associated with the packaging, containerization, blocking and bracing, ventilation, stability of shipping, stacking and storage configuration.
duty time [STD 8719.24 Annex]	the time personnel are at work from the time they arrive at their duty location until the end of the duty tour; duty time begins on first arriving at the base or office for transportation to later launch support positions.
Earth Ascent Abort [NPR 8705.2]	An abort performed during Earth ascent, where the crewed spacecraft is separated from the launch vehicle without the capability to achieve a safe stable orbit. The crew is safely returned to Earth in a portion of the spacecraft nominally used for entry and landing/touchdown.
Earth-Covered Magazine (ECM) [STD 8719.12]	An aboveground, earth-covered structure intended for the storage of explosives, pyrotechnics, propellant, or United Nations (UN) Class 1 hazardous materials that meets soil cover depth and slope requirements of this standard.
Eastern and Western Range 127-1 [STD 8719.24 Annex]	Eastern and Western Range 127-1, Range Safety Requirements refers to the previous Range Safety requirements directive that controlled range and Range User activities at the Eastern and Western Ranges.
Eastern Range [STD 8719.24 Annex]	part of the National Launch Range facilities, operated by the 45th Space Wing, part of Air Force Space Command, and located at Patrick Air Force Base, Florida; the range includes the operational launch and base support facilities located at Cape Canaveral Air Force Station, Florida, radar tracking sites and ground stations located in the eastern Caribbean as well as the Jonathan-Dickson Missile Tracking Annex (Jupiter, Florida) and Argentia, Newfoundland sites.
Eccentricity [STD 8719.14]	The apogee altitude minus perigee altitude of an orbit divided by twice the semi major axis. Eccentricity is zero for circular orbits and less than one for all elliptical orbits.
Eddy Current Brake [STD 8719.9]	An electrical induction brake used to reduce or control speed.
Egress [STD 8719.11]	A continuous and unobstructed way of travel from any point in a building or structure to a public way. It consists of three separate and distinct parts (a) the exit access, (b) the exit, and (c) the exit discharge. A means of egress comprises the vertical and horizontal ways of travel and includes intervening room spaces, doorways, hallways, corridors, passageways, ramps, stairs, lobbies, horizontal exits, courts, and sidewalks.
electrical component [STD 8719.24 Annex]	a component such as a switch, fuse, resistor, wire, capacitor, or diode in an electrical system.

Term [Citing Document(s)]	Definitions
Electro-Explosive Device (EED) [STD 8719.12]	An electrically-initiated device containing an explosive or pyrotechnic mixture. The output of the initiation is heat, shock, or mechanical action.
Electromagnetic Interference [STD 8739.4]	The unwanted intrusion of electromagnetic radiation energy whose frequency spectrum extends from subsonic frequency to X-rays.
Electrostatic Discharge (ESD) [STD 8719.12]	The rapid and spontaneous transfer of electrical charge between two bodies at different electrical potentials.
Element [STD 8739.9]	The generic term applied to the smallest portions of a software or document product that can be independently developed or modified.
ELV Payload Safety Agency Team [NPR 8715.7]	An Agency group appointed by the Chief, Safety and Mission Assurance that performs as an element of the NASA OSMA and provides guidance to the NASA Chief, Safety and Mission Assurance, the NASA ELV Payload Safety Manager, and NASA ELV payload projects. The Agency Team works with the Payload Safety Working Group to resolve any safety concerns associated with a project. The Agency Team also works to ensure that NASA ELV payload safety policy and requirements are adequate and consistently implemented throughout the Agency.
ELV Payload Safety Manager [NPR 8715.7]	A position appointed by the Chief, Safety and Mission Assurance that leads the ELV Payload Safety Program, ensuring Agency policy, requirements, and processes are developed, maintained, and implemented to safeguard people and resources from hazards associated with payload to launch vehicle integration, multiple payloads, and payloads and related GSE. This individual also leads the Agency Team.
Emergency [NPR 8715.3]	Unintended circumstance bearing clear and present danger to personnel or property which requires an immediate response.
Emergency Egress [NPR 8705.2]	Capability for a crew and passengers to exit the vehicle and leave the hazardous situation or catastrophic event within the specified time. Crew/passenger emergency egress can be unassisted or assisted by ground personnel.
Emergency Egress [NPR 8715.3]	The capability for an unassisted crew to exit a vehicle and leave a hazardous situation within a specified amount of time.
Emergency Equipment and Systems [NPR 8705.2]	A set of components (hardware and/or software) used to mitigate or control hazards, after occurrence, which present an immediate threat to the crew or crewed spacecraft. Examples include fire suppression systems and extinguishers, emergency breathing devices, and crew escape systems.
Emergency Medical [NPR 8705.2]	The capability to respond to crew illness or injury in order to prevent, or mitigate, crew demise or permanent disability. This includes either an inherent capability on a vehicle, timely transfer to a place or vehicle that can provide a higher level of medical care, or both.
Emergency Medical [NPR 8715.3]	The capability to respond to illness or injury in order to prevent fatality or permanent disability. This capability includes either an inherent local capability or the timely transfer to a place or vehicle that can provide a similar or higher level of medical care, or both.

Term [Citing Document(s)]	Definitions
Emergency Response Planning Guidelines - Level 2 [STD 8719.25]	The Emergency Response Planning Guidelines - Level 2 is the maximum airborne concentration below which it is believed nearly all individuals could be exposed for up to one hour without experiencing or developing irreversible or other serious health effects or symptoms that could impair an individual's ability to take protective action.
Emergency Stop (E-Stop) [STD 8719.9]	A manually operated switch or valve to cut off electric or fluid power independently of the regular operating controls.
Emergency Systems [NPR 8715.3]	A set of components (hardware and/or software) used to mitigate or control hazards which present an immediate threat to the crew or crewed spacecraft. Examples include fire suppression systems and extinguishers, emergency breathing devices, and crew escape systems.
Encapsulation [STD 8739.1]	The complete encasement of a component or module in a resin. Other terms used in the electronics industry to indicate encapsulation are "potting," "embedment," and "molding."
End of Period Reliability [STD 8739.12]	The probability of an item of measuring and test equipment being in tolerance at the end of its assigned calibration interval.
Endorsing Official [NPR 8621.1]	An official who reviews the signed mishap investigation report and provides a signed written endorsement, comments, and when not the appointing official, a recommendation for the report approval or rejection by the appointing official.
Energetic Liquid [STD 8719.12]	A liquid, slurry, or gel, consisting of, or containing, an explosive, oxidizer, fuel, or combination of the above, that may undergo, contribute to, or cause rapid exothermic decomposition, deflagration, or detonation.
Energetic Material [STD 8719.12]	A material consisting of, or containing, an explosive, oxidizer, fuel, or combination of the above, that may undergo, contribute to, or cause rapid exothermic decomposition, deflagration, or detonation.
Enhanced Low Dose Rate Sensitivity (ELDRS) [STD 8739.10]	The characteristic of a device that exhibits an enhanced total dose response at dose rates below 50 rad(Si)/s.
Entry / Entry Operation [NPR 8715.5]	The sequence of controlled thrust maneuvers or other events that brings a space vehicle or spacecraft from Earth orbit or outer space to Earth. Entry operations do not include suborbital flights.
Entry Operation [STD 8719.25]	The sequence of controlled thrust maneuvers or other events that brings a space vehicle or spacecraft from Earth orbit or outer space to Earth. Entry operations do not include suborbital flights.
Environment [STD 8739.9]	The components and features that are not part of the product but necessary for its execution such as software, hardware, and tools. (see JSC 31011)
Environment [STD 8729.1]	The natural and induced conditions experienced by a system including its people, processes, and products during operational use, stand-by, maintenance, transportation, and storage.

Term [Citing Document(s)]	Definitions
Environmental Health [STD 8719.24 Annex]	on the Western Range, the Range User is responsible for performing the EH tasks described in this document for contractor operations; on the Eastern Range, the responsible agency is 45 MG/SGPB and a range contractor.
Equivalent Entity [STD 8719.9]	A person or organization (including an employer) which by possession of equipment, technical knowledge and skills, can perform with equal competence the same repairs and tests as the person or organization with which it is equated.
Equivalent Level of Safety [NPR 8715.7]	An alternate approach to meet the intent of a requirement that provides no additional risk as determined by qualitative or quantitative means.
equivalent level of safety [STD 8719.24 Annex]	an approximately equal level of safety; may involve a change to the level of expected risk that is not statistically or mathematically significant as determined by qualitative or quantitative risk analysis; equivalent level of safety replaces the former "meets intent" certification process.
Equivalent Level of Safety (ELS) (determination) [NPR 8715.5]	The approval of an alternative approach to satisfying a range flight safety requirement where the alternative provides an approximately equal level of safety as determined by qualitative or quantitative means.
Equivalent Level of Safety (ELS) (determination) [STD 8719.25]	The approval of an alternative approach to satisfying a range safety requirement where the alternative provides an approximately equal level of safety as determined by qualitative or quantitative means.
Equivalent/Equivalency [STD 8719.11]	When referring to fire protection and life safety, the technology, systems, devices, and designs that, while not meeting the letter of code provisions, will provide comparable levels of fire safety. This determination is to be made by the AHJ after a complete analysis of hazardous conditions and required levels of safety.
Error [STD 8739.9]	The difference between a computed, observed, or measured value or condition and the true, specified, or theoretically correct value or condition. (see IEEE Std 610.12-1990)
Escape [NPR 8705.2]	Removal of crew and passengers from the portion of the space system normally used for reentry, due to rapidly deteriorating and hazardous conditions, thus, placing them in a safe situation suitable for survivable return or recovery. Escape includes, but is not limited to, those modes that utilize a portion of the original space system for the removal (e.g., pods, modules, or fore bodies).

Term [Citing Document(s)]	Definitions
Essential Electronic Equipment [STD 8719.11]	Equipment that meets one or more of the following criteria: • Is directly related to the NASA mission and which, if lost, would seriously impact the ability of NASA to perform its mission. • Is necessary to the safety and health of personnel. • Is essential to the security or health of the Nation. • Performs an operation that must be continued to completion without termination. • Performs an operation which could be performed by substitute methods, but where the substitute methods would involve significant additional expenditures for personnel, facilities, and/or equipment or an unacceptable length of time. • Has a high monetary value to the Federal Government (greater than $1 million). Electronic equipment includes all equipment and devices that are electrically powered and use the emission of electrons in vacuum tubes, cathode ray tubes, photoelectric cells, transistors, diodes, integrated circuits, and other solid state devices. This includes, but is not limited to, electronic digital and analog computers, telephone communications and switching equipment, and other electronic equipment used for statistics, communication, process control, measurement, guidance, simulation, or supervisory operations.
Event [NPR 8621.1]	A real-time occurrence describing one discrete action, typically an error, failure, or malfunction (e.g., pipe broke, power lost, lightning struck, and person opened valve).
Event and Causal Factor Tree [NPR 8621.1]	A graphic representation of the mishap or close call that shows the event (accident) at the top of the tree; depicts the logical sequence of events; illustrates all causal factors (including conditions and failed barriers) necessary and sufficient for the mishap or close call occurrence; and depicts the root causes at the bottom of the tree.
Evidence [NPR 8621.1]	Everything used to support or refute a hypothesis or finding. For a safety investigation, the types of evidence are physical (e.g., hardware), demonstrable (24 hours in one day), witness interview, and documentary (witness statement, logbooks, and electronic data).
Ex Officio [NPR 8621.1]	An individual tasked to ensure the investigation conducted conforms to NASA policy and this NPR.
Exception [NPR 8705.2]	A written authorization granting relief from a specific, non-applicable requirement. NPR 7120.5 defines non-applicable requirement as "Any requirement not relevant; not capable of being applied." The term exception is generally no longer used. For the purposes of this NPR, the term "exception" is equivalent to and interchangeable with a "Determination of nonapplicability" as described in NPR 8715.3.
Exception [NPR 8715.3]	An authorization for permanent relief from a specific requirement and may be requested at any time during the life cycle of a program/project.
Exception [STD 8709.20]	The term "exception" is no longer used. "Non-applicable" requirement determination is made via a deviation or waiver request.
Excluded PVS [STD 8719.17]	A PVS that is not required to meet the certification (or recertification) requirements of NPD 8710.5, Policy for Pressure Vessels and Pressurized Systems, and need not be included in the PVS configuration management system except as noted in specific exclusion clauses of this document. Excluded PVS are subject to other applicable laws, regulations, safety requirements, NASA requirements, and appropriate VCS and must be maintained in accordance with applicable VCS.

Term [Citing Document(s)]	Definitions
Executive Summary [NPR 8621.1]	A top-level summary, which is part of the mishap investigation report, describing the circumstances of a mishap including who, what, when, where, and why, and a description of the proximate and root causes. The executive summary should be worded where possible to meet NASA's Office of Communications criteria for public release.
Exemption [NPR 8705.2]	A written authorization granting relief from the space system failure tolerance requirement.
Existing PVS [STD 8719.17]	PVS are considered to be "Existing PVS" if installed no later than 6 months from the date of original issue of this document (9/22/2006).
Expectation of Casualty (Ec) [STD 8719.25]	The average number of casualties expected per an event, such as vehicle flight, if a large number of events could be carried out under identical circumstances.
Expendable Launch Vehicle [STD 8719.25]	A vehicle that, once launched, is not reused and typically is not retrieved.
expendable launch vehicle [STD 8719.24 Annex]	a vehicle that, once launched, is typically not retrieved and reused.
Expendable Launch Vehicle (ELV) [NPR 8715.5]	A vehicle that, once launched, is not reused and typically is not retrieved.
explosion proof apparatus [STD 8719.24 Annex]	an enclosure that will withstand an internal explosion of gases or vapors and prevent those gases or vapors from igniting the flammable atmosphere surrounding the enclosure, and whose external temperature will not ignite the surrounding flammable atmosphere.
Explosive Debris [STD 8719.25]	Solid propellant fragments or other pieces of a launch or entry vehicle or payload that result from breakup of the vehicle during flight and could explode upon impact with the Earth's surface or on their own.
Explosive Donor [STD 8719.12]	An explosion from a small device or explosive mass that may cause an adjacent explosive item or larger mass to react to yield measurable blast overpressure.
Explosive Equivalent (TNT Equivalent) [STD 8719.12]	Amount of a standard explosive that, when detonated, will produce a blast effect comparable to that which results at the same distances from the detonation or explosion of a given amount of the material for which performance is being evaluated. It usually is expressed as a percentage of the total net weight of all reactive materials contained in the item or system. For the purpose of these standards, TNT is used for comparison.
Explosive(s) [STD 8719.12]	Any chemical compound or mechanical mixture that, when subjected to heat, impact, friction, detonation, or other suitable initiation, undergoes a very rapid chemical change with the evolution of large volumes of highly heated gases that exert pressures in the surrounding medium. The term applies to materials that either detonate or deflagrate.

Term [Citing Document(s)]	Definitions
explosives [STD 8719.24 Annex]	any chemical compound or mechanical mixture that, when subjected to heat, impact, friction, detonation, or other suitable initiation, undergoes a very rapid chemical change with the evolution of large volumes of highly heated gases that exert pressures in the surrounding medium; the term applies to materials that either detonate or deflagrate.
Explosives Area [STD 8719.12]	A restricted area specifically designated and set aside from other portions of a Center for the manufacturing, processing, storing, and handling of explosives.
Explosives Handler [STD 8719.12]	Certified personnel authorized to physically handle explosives or pyrotechnic devices (outside of transportation or packing configuration) during storage, installation, inspection, or other use identified in an approved procedure.
Exposed Explosives [STD 8719.12]	Explosives that are open to the atmosphere (such as unpackaged bulk explosives or disassembled or open components) and susceptible to initiation directly by static or mechanical spark, or create (or accidentally create) explosive dust, or give off vapors, fumes, or gases in explosive concentrations. This also includes exudation and explosives exposed from damaged items such as gun powder or rocket motors.
Exposed Site (ES) [STD 8719.12]	A location exposed to the potential hazardous effects (blast, fragments, debris, and heat flux) from an explosion at a potential explosion site (PES). The distance to a PES and the level of protection required for an ES determine the quantity of explosives permitted in a PES.
Exposure [NPR 8715.3]	(1) Vulnerability of a population, property, or other value system to a given activity or hazard; or (2) other measure of the opportunity for failure or mishap events to occur.
f10 [STD 8719.14]	An index of solar activity; often a 13-month running average of the energy flux from the Sun measured at a wavelength of 10.7 cm, expressed in units of 104 janskys.
Facility [STD 8719.11]	Buildings, structures, and other real property improvements including utilities and collateral equipment.
factor of safety [STD 8719.24 Annex]	the ratio of the yield or ultimate strength of the structure to the applied load; see factor of safety (ultimate) and factor of safety (yield); ratio of the design condition to the maximum operating conditions specified during design.
Factor of Safety (FS) [STD 8719.17]	Unless otherwise noted, this refers to the material design factor of safety on structural failure and is equal to the lesser of the material strength divided by the material stress under anticipated loading or the actual buckling load divided by the anticipated buckling load.
Factor of Safety (Safety Factor) [NPR 8715.3]	Ratio of the design condition to the maximum operating conditions specified during design (see also Safety Margin and Margin of Safety).
Fail Safe [STD 6008]	A fracture control classification based on redundancy where, after failure of a single fastener, the remaining structure can withstand the redistributed loads and the failure will not release a potentially catastrophic free body. A fail safe fastener meets the criteria specified in NASA-STD-5019, section 4.1.1.3.

Term [Citing Document(s)]	Definitions
Fail-Safe [NPR 8715.3]	Ability to sustain a failure and retain the capability to safely terminate or control the operation.
fail-safe [STD 8719.24 Annex]	a design feature in which a system reacts to a failure by switching to or maintaining a safe operating mode that may include system shut down; ability to sustain a failure and retain the capability to safely terminate or control the operation.
Failure [NPR 8705.2]	Inability of a system, subsystem, component, or part to perform its required function within specified limits (Source - NPR 8715.3).
Failure [NPR 8715.3]	Inability of a system, subsystem, component, or part to perform its required function within specified limits.
Failure [STD 8739.9]	The behavior of the software or system component when a fault is encountered, producing an incorrect or undesired effect of a specified severity.
Failure [STD 8729.1]	[1] Inability of a system, subsystem, component, or part to perform its required function within specified limits. [2] Non-performance or incorrect performance of an intended function of a product. A failure is often the manifestation of one or more faults and is permanent.
failure [STD 8719.24 Annex]	the inability of a system, subsystem, component, or part to perform a required function within specified limits.
Failure Analysis [STD 8729.1]	The conduct of evaluations and analyses to determine the specific cause of system (including elements of hardware, software, and human performance) and/or component failure.
Failure Cause [STD 8729.1]	The defect in design, process, quality, or part application that is the underlying cause of a failure or which initiates a process that leads to failure.
Failure Effect [STD 8729.1]	The immediate consequence of a failure on operation, function or functionality.
Failure Mechanism [STD 8729.1]	The process (e.g., physical, chemical, electrical, thermal) of degradation or the chain of events, which results in a particular failure mode.
Failure Mode [NPR 8715.3]	Particular way in which a failure can occur, independent of the reason for failure.
Failure Mode [STD 8729.1]	[1] Particular way in which a failure can occur, independent of the reason for failure. [2] The characteristic manner in which a failure occurs, independent of the reason for failure; the condition or state that is the end result of a particular failure mechanism; the consequence of the failure mechanism through which the failure occurs, e.g., short, open, fracture, excessive wear.
Failure Mode Effects and Criticality Analysis [STD 8739.10]	Analysis of a system and the working interrelationships of its elements to determine ways in which failures can occur (failure modes) and the effects of each potential failure on the system element in which it occurs, on other system elements, on the mission, and the study of the relative mission significance or criticality of all potential failure modes.

Term [Citing Document(s)]	Definitions
Failure Mode Effects and Criticality Analysis (FMECA) [STD 8729.1]	Analysis of a system and the working interrelationships of its elements to determine ways in which failures can occur (failure modes) and the effects of each potential failure on the system element in which it occurs, on other system elements, and on the mission, and the study of the relative mission risk or criticality of all potential failure modes.
Failure Modes and Effects Analysis (FMEA) [NPR 8715.3]	A bottoms up systematic, inductive, methodical analysis performed to identify and document all identifiable failure modes at a prescribed level and to specify the resultant effect of the modes of failure. It is usually performed to identify critical single failure points in hardware. In relation to formal hazard analyses, FMEA is a subsidiary analysis.
Failure Modes and Effects Analysis (FMEA) [STD 8719.9]	A systematic, methodical analysis performed to identify and document failure modes and their resultant effects at a prescribed level.
Failure Modes and Effects Analysis (FMEA) [STD 8729.1]	[1] A bottoms up systematic, inductive, methodical analysis performed to identify and document all identifiable failure modes at a prescribed level and to specify the resultant effect of the modes of failure. It is usually performed to identify critical single failure points in hardware. In relation to formal hazard analyses, FMEA is a subsidiary analysis. [2] A bottom-up systematic, inductive, methodical analysis performed to identify and document all identifiable failure modes at a prescribed level and to specify the resultant effect of the modes of failure. [3] Analysis of a system and the working interrelationships of its elements to determine ways in which failures can occur (failure modes) and the effects of each potential failure on the system element in which it occurs, on other system elements, and on the mission.
Failure Propagation [STD 8729.1]	Any physical or logical event caused by failure within a product which can lead to failure(s) of products outside the boundaries of the product under analysis.
Failure Tolerance [NPR 8705.2]	The ability to sustain a certain number of failures and still retain capability.
Failure Tolerance [NPR 8715.3]	Built-in capability of a system to perform as intended in the presence of specified hardware or software failures.
Failure Tolerance [STD 8729.1]	The ability to perform a function in the presence of any of a specified number of coincident, independent failure causes of specified types.
failure, ductile [STD 8719.24 Annex]	materials exhibiting a ductile failure mode are those that (1) have ductile behavior under the environmental and operating conditions; i.e., ultimate strain of 20 percent elongation or greater, and appropriate notch toughness, and (2) provide warning of an incoming failure via visually detectable (by eye and without magnification aids) deformation of structural components; see also ductile behavior.
Faraday Cap [STD 8719.12]	A conductive metal cap that can be placed over the connector of an Electro-explosive Device (EED), e.g., an NASA Standard Initiator (NSI), to prevent inadvertent firing from exposure to an external electric field, i.e., radio frequency (RF) sources. Some, but not all, Faraday caps also short the bridgewire.

Term [Citing Document(s)]	Definitions
Fastener [STD 6008]	An item such as a bolt (could be a tensile or shear bolt, shoulder bolt, screw, HiLok®, HiTigue®, or lockbolt), nut, nut plate or anchor nut, rivet, shear pin, helical or cylindrical insert, setscrew, washer, safety wire, cotter pin, etc., which joins or retains components or structural elements.
fatigue [STD 8719.24 Annex]	the progressive localized permanent structural change that occurs in a material subjected to constant or variable amplitude loads at stresses having a maximum value less than the ultimate strength of the material.
fatigue life [STD 8719.24 Annex]	the number of cycles of stress or strain of a specified character that a given material sustains before failure of a specified nature occurs.
Fault [NPR 8705.2]	An undesired system state and/or the immediate cause of failure (e.g., maladjustment, misalignment, defect, or other). The definition of the term "fault" envelopes the word "failure," since faults include other undesired events such as software anomalies and operational anomalies (Source - MIL-STD-721C). Faults at a lower level could lead to failures at the higher subsystem or system level.
Fault [STD 8739.9]	A manifestation of an error in software. If encountered, a fault may cause a failure.
Fault [STD 8729.1]	[1] An undesired system state and/or the immediate cause of failure (e.g., maladjustment, misalignment, defect, or other). The definition of the term "fault" envelopes the word "failure," since faults include other undesired events such as software anomalies and operational anomalies. [2] An inherent defect in a product which may or may not ever manifest, such as a bug in software code.
fault [STD 8719.24 Annex]	the manifestation of an error in software that may cause a failure.
Fault Detection [STD 8739.9]	The ability to perform checks to determine whether any erroneous situation has arisen.
Fault Isolation [STD 8729.1]	The process of determining the approximate location of a fault.
Fault Management [STD 8729.1]	The engineering process that encompasses practices which enable an operational system to contain, prevent, detect, isolate, diagnose, respond to, and recover from conditions that may interfere with nominal mission operations.
Fault Propagation [STD 8729.1]	The propagation of effects seen from one fault into other faults and potentially failures.
Fault Recovery [STD 8739.9]	The response of the system software to an abnormal condition, so that system execution can continue to yield correct results despite the existence of the fault.
fault tolerance [STD 8719.24 Annex]	the built-in ability of a system to provide continued correct operation in the presence of a specified number of faults or failures.

Term [Citing Document(s)]	Definitions
Fault Tree [NPR 8715.3]	A schematic representation resembling an inverted tree that depicts possible sequential events (failures) that may proceed from discrete credible failures to a single undesired final event (failure). A fault tree is created retrogressively from the final event by deductive logic.
Fault Tree Analysis [NPR 8621.1]	An analytical technique whereby an undesired system state is specified, and the system is then analyzed in the context of its environment and operation to find all credible ways in which the undesired event can occur.
Fault Tree Analysis [NPR 8715.3]	An analysis that begins with the definition or identification of an undesired event (failure). The fault tree is a symbolic logic diagram showing the cause-effect relationship between a top undesired event (failure) and one or more contributing causes. It is a type of logic tree that is developed by deductive logic from a top undesired event to all sub-events that must occur to cause it.
Fault Tree Analysis [STD 8739.10]	A deductive system reliability tool that provides both qualitative and quantitative measures of the probability of failure. It estimates the probability that a top-level event will occur, systematically identifies all possible causes leading to the top event, and documents the analytic process to provide a baseline for future studies of alternative designs.
Fault Tree Analysis (FTA) [STD 8729.1]	A deductive system reliability tool that provides both qualitative and quantitative measures of the probability of failure. It estimates the probability that a top-level event will occur, systematically identifies all possible causes leading to the top event, and documents the analytic process to provide a baseline for future studies of alternative designs.
Federal employee [NPR 8621.1]	(Per 5 U.S.C. pt. 2101) a. Civil service consists of all appointive positions in the executive, judicial, and legislative branches of the Government of the United States, except positions in the uniformed services. b. Armed forces means the Army, Navy, Air Force, Marine Corps, and Coast Guard. c. Uniformed services means the armed forces, the commissioned corps of the Public Health Service, and the commissioned corps of the National Oceanic and Atmospheric Administration.
Ferrule [STD 8739.4]	A short metal tube used to make crimp connections to shielded or coaxial cables.
Ferrule [STD 8739.5]	A mechanical fixture, generally a rigid tube, used to confine the stripped end of a fiber bundle or an optical fiber.
Fiber (Optical) [STD 8739.5]	A filament shaped optical waveguide made of dielectric material.
Fiber Optic Cable [STD 8739.5]	A fiber, multiple fiber or fiber bundle in a cable structure fabricated to meet optical mechanical and environmental specifications.
Fiber Optic Connector [STD 8739.5]	A fiber optic component normally assembled onto a cable and attached to a piece of apparatus for the purpose of providing interconnecting/disconnecting of fiber optic cables.
Filler [STD 8739.1]	A material added to polymers in order to reduce cost or modify physical properties.

Term [Citing Document(s)]	Definitions
Fillet [STD 8739.1]	A smooth, generally concave, buildup of material between two surfaces (e.g., a buildup of conformal coating material between a part and the printed circuit board (PCB)).
Fillet [STD 8739.4]	A smooth concave buildup of material between two surfaces; e.g., a fillet of solder between a conductor and a solder terminal.
Final Acceptance [NPR 8735.2]	The act of an authorized representative of the Government by which the Government, for itself or as an agent of another, assumes ownership of existing identified supplies tendered or approves specific services rendered as partial or complete performance of the contract.
Final Mishap Investigation Report [NPR 8621.1]	The signed mishap investigation report with endorsements and comments attached.
Finding [NPR 8705.6]	A conclusion based on facts and objective evidence or lack thereof established during SMA audits, reviews, and assessments.
Finding [NPR 8621.1]	A conclusion, positive or negative, based on facts established by the investigating authority during the investigation (i.e., cause, contributing factor, and observation).
Fire Partition [STD 8719.11]	A physical barrier to prevent the horizontal spread of fire between areas within buildings, constructed of materials sufficient to achieve a 1- or 2-hour fire-resistance rating as determined by NFPA 251. The barrier must extend from the floor to the floor/roof above the area involved (partitions may extend to a listed membrane ceiling at the discretion of the AHJ). Large openings in partitions must be protected by listed fire doors or fire dampers. "Pokethrough" openings must be sealed with noncombustible materials listed for that use. Fire partitions are not to be confused with fire walls which have a greater hourly fire resistance and are capable of independent support. (See definition of firewall.)
Fire Wall [STD 8719.11]	A physical barrier to prevent the horizontal spread of fire between buildings, constructed of materials sufficient to achieve at least a 3 or 4 hour fire resistance rating as determined by NFPA 251.
Firebrand [STD 8719.12]	A projected burning or hot fragment whose thermal energy has the potential for transfer to a receptor.
Fire-Resistive [STD 8719.11]	A broad range of structural systems capable of withstanding maximum intensity and duration of fire without failure. Common fire-resistive components include masonry load-bearing walls, reinforced concrete or protected steel columns, and poured or pre-cast concrete floors and roofs.
firing circuit [STD 8719.24 Annex]	the current path between the power source and the initiating device.
Firmware [STD 8719.13]	The combination of a hardware device and computer instructions and data that reside as read-only software on that device. This term is sometimes used to refer only to the hardware device or only to the computer instructions or data, but these meanings are deprecated. The confusion surrounding this term has led some to suggest that it be avoided altogether. For the purposes of this Standard Firmware is considered as software. Firmware is NOT the same as Programmable Logic Devices/Complex Electronics.

Term [Citing Document(s)]	Definitions
Firmware [STD 8739.9]	The combination of a hardware device and computer instructions and/or data that reside as read-only software on that device. This term is sometimes used to refer only to the hardware device or only to the computer instructions or data, but these meanings are deprecated. The confusion surrounding this term has led some to suggest that it be avoided altogether. For the purposes of this Standard Firmware is considered as software. Firmware is NOT the same as Programmable Logic Devices/Complex Electronics.
firmware [STD 8719.24 Annex]	computer programs and data loaded in a class of memory that cannot be dynamically modified by the computer during processing; for Systems Safety purposes, firmware is to be treated as software.
First Aid [NPR 8621.1]	Refer to OSHA definition in 29 CFR pt. 1904.7.
fittings [STD 8719.24 Annex]	pressure components of a pressurized system initialized to connect lines, other pressure components, and/or pressure vessels within the system.
Flammable Liquid [STD 8719.12]	Any liquid having a flash point below 100 °F (38 °C) and a vapor pressure not exceeding 280 kPa (41 psia) at 100 °F (37.8 °C). This is the definition as applied in this manual; it includes some materials defined as combustible liquids by the Department of Transportation (DOT).
Flammable Liquid [STD 8719.11]	A liquid having a flash point below 100 °F (37.9 °C) and having a vapor pressure not exceeding 40 pounds per square inch (absolute (275.79 kilopascal) at I00 °F (37.9° C)) or a combustible liquid heated to, or above, its flash point.
Flash Point [STD 8719.12]	The lowest temperature of the test specimen, adjusted to account for variations in atmospheric pressure from 101.3 kPa, at which application of an ignition source causes the vapors of the test sample to ignite under specified conditions of test.
Flatpack [STD 8739.1]	A part with two straight rows of leads (normally on 1.27mm (0.050 inch) centers) that are parallel to the part body.
flaw [STD 8719.24 Annex]	an imperfection or unintentional discontinuity that is detectable by nondestructive examination.
Flexible Hose [STD 8719.17]	A non-rigid piping component excluding bellows expansion joints.
Flight [NPR 8715.5]	Launch or entry of an orbital or suborbital space vehicle/spacecraft or operation of an aeronautical vehicle (to include aircraft, UAS, and balloons). For the purposes of this NPR, "flight" does not include on-orbit or interplanetary operations.
Flight [STD 8719.25]	Launch or entry of an orbital or suborbital space vehicle/spacecraft or operation of an aeronautical vehicle (to include aircraft, UAS, and balloons). For the purposes of this standard, "flight" does not include on-orbit operations.
Flight Hardware [NPR 8621.1]	Any hardware that is flown on or part of an aircraft, experimental flight vehicle, satellite, lighter than air vehicles, unoccupied aerial vehicle, or space transportation system.
Flight Hardware	Hardware designed and fabricated for ultimate use in a vehicle intended to fly.

Term [Citing Document(s)]	Definitions
[NPR 8715.3] [NPR 8715.7]	
flight hazard area [STD 8719.24 Annex]	a hazardous launch area; the controlled surface area and airspace about the launch pad and flight azimuth where individual risk from a malfunction during the early phase of flight exceeds 1 x 10-5; because the risk of serious injury or death from blast overpressure or debris is so significant, only launch-essential personnel in approved blast-hardened structures with adequate breathing protection are permitted in this area during launch.
flight plan approval [STD 8719.24 Annex]	an approval process that results from a written application by the Range User; a two-phase approach stemming from a Preliminary Flight Plan Approval and a Final Flight Plan Approval.
Flight PVS [STD 8719.17]	An assembly of components under pressure, including vessels, piping, valves, relief devices, pumps, expansion joints, gages, etc., that are fabricated in accordance with program requirements specifically for use in aircraft or spacecraft.
Flight Safety Analyst [STD 8719.25]	A person responsible for identifying and analyzing all hazards to people and property associated with flight operations, e.g. debris, toxics, DFO, and COLA, through qualitative and quantitative methods. A flight safety analyst performs risk assessments on the flight operation design to determine risk levels and support risk management efforts to ensure criteria are met by establishing any design or operational constraints needed to control hazards and risks to people and property.
Flight Safety Officer (FSO) [NPR 8715.5]	A person responsible for safety during a range flight operation. An FSO has the authority to hold or abort the operation, or take a risk mitigation action, which includes terminating the flight. FSO is synonymous with the term Mission Flight Control Officer (MFCO) used at some DoD ranges.
Flight Safety Officer (FSO) [STD 8719.25]	A person responsible for real-time safety during a range flight operation. An FSO has the authority to hold or abort the operation, or take a risk mitigation action, which includes terminating the flight. FSO is synonymous with the term MFCO used at some DoD ranges.
Flight Safety System Engineer [STD 8719.25]	A person responsible for ensuring that the flight safety system (which includes the Flight Termination System (FTS) or Contingency Management System (CMS), and associated tracking and telemetry systems) is designed, qualified, and operated in accordance with applicable requirements or standards.
Flight Safety System(s) (FSS) [NPR 8715.5] [STD 8719.25]	A system (including any subsystem) whose performance is factored into the Range Safety Analysis and relied upon during flight to mitigate hazards. These systems include range safety displays, range clearance capability, radar, optic tracking systems, telemetry, tracking display systems (including instantaneous impact predictors), contingency management systems, flight termination systems, and command and control capability for flight termination systems.
Flight Software [NPR 8621.1]	Any software that is flown on or part of an aircraft, experimental flight vehicle, satellite, lighter than air vehicles, unoccupied aerial vehicle, or space transportation system.

Term [Citing Document(s)]	Definitions
flight termination system [STD 8719.24 Annex]	all components, onboard a launch vehicle, that provide the ability to terminate a launch vehicle's flight in a controlled manner; the flight termination system consists of all command terminate systems, inadvertent separation destruct systems, or other systems or components that are onboard a launch vehicle and used to terminate flight.
Flight Termination System (FTS) [NPR 8715.5]	A type of Flight Safety System designed, tested, and incorporated into vehicles that provides for the independent and deliberate termination of an errant/erratic vehicle's flight.
Flight Termination System (FTS) [STD 8719.25]	A type of Range Safety System designed, tested, and incorporated into vehicles that provides for the independent and deliberate termination of an errant/erratic vehicle's flight.
Flow down [NPR 8705.6]	The documented demonstration that the Center is operating in accordance with the Agency requirements through references to either each requirement of the Agency documents, or a general "shall" statement denoting the Agency document within the Center document. Flow down is accomplished by a Center document "shall" statement that invokes the Agency requirement, a generic shall statement identifying the Agency document within the Center document, references to each Agency requirement in the Center document, or a more stringent, Center-specific requirement identified in the Center document.
Flux [STD 8739.1]	A chemically-active compound that, when heated, removes minor surface oxidation, minimizes oxidation of the basis metal, and promotes the formation of an intermetallic layer between solder and basis metal.
foreign government agency or company [STD 8719.24 Annex]	a Range User entity who is not a US citizen, not a US company, or not a foreign-registered company with a majority holding by a US company or citizen.
Formal Inspection [STD 8739.9]	A set of practices used to perform inspections in a precise, repeatable manner which includes the following specific Inspection process steps: (1) Planning, (2) Overview, (3) Preparation, (4) Inspection meeting, (5) Rework, and (5) Follow-up. It also has built in self-improvement process which includes the collection of data with which one can analyze the effectiveness of the process and track and make changes.
fracture control [STD 8719.24 Annex]	the application of design philosophy, analysis method, manufacturing technology, quality assurance, and operating procedures to prevent premature structural failure due to the propagation of cracks or crack-like flaws during fabrication, testing, transportation and handling, and service.
fracture mechanics [STD 8719.24 Annex]	an engineering concept used to predict flaw growth of materials and structures containing cracks or crack-like flaws; an essential part of a fracture control plan to prevent structure failure due to flaw propagation.
fracture toughness [STD 8719.24 Annex]	a generic term for measures of resistance to extension of a crack.
fracture, brittle [STD 8719.24 Annex]	for the purpose of this document, those materials that exhibit a failure mode outside of ductile failure.

Term [Citing Document(s)]	Definitions
Fracture-Critical Fastener [STD 6008]	A classification that assumes that fracture or failure of the fastener resulting from the occurrence of a crack will result in a catastrophic hazard, as specified in NASA-STD-5019.
Fragmentation [STD 8719.12]	Breaking up of the confining material of a chemical compound or mechanical mixture when an explosion takes place. Fragments may be complete items, subassemblies, pieces thereof, or pieces of equipment or buildings containing the items.
Franchised Distributor [STD 8739.10]	A source authorized by the original component manufacturer to distribute parts.
Free Space Environment [STD 8739.10]	The natural space radiation environment present in the absence of any man-made structures or objects. This definition only applies above the Kármán Line (100 km altitude).
FTS Command System [STD 8719.25]	All components needed to send a flight termination command signal to an onboard vehicle flight termination system. An FTS command system starts with flight termination activation controls and ends at each command-transmitting antenna. It includes all intermediate equipment linkages, software, and auxiliary transmitters that ensure a command signal will reach the onboard vehicle flight termination system during flight.
Fuel Load (a.k.a. Fire Load) [STD 8719.11]	Expected maximum quantity of combustible material in a given fire area. In normal facilities, the combustible structural elements and the combustible contents contained within that area. Fire load is usually expressed as weight of combustible material per square foot of area.
function [STD 8719.24 Annex]	any electronic commands, such as arm, destruct, safe, and test, issued by the Mission Flight Control Officer and transmitted to the airborne elements of a flight termination system.
Functional Configuration Audit (FCA) [STD 8739.8]	An audit conducted to verify that the development of a configuration item has been completed satisfactorily, that the item has achieved the performance and functional characteristics specified in the functional or allocated configuration identification, and that its operational and support documents are complete and satisfactory.
Functional Failure [NPR 8735.1]	A failure resulting in nonfulfillment of required component functions or capabilities.
Functional Redundancy [NPR 8715.3]	A situation where a dissimilar device provides safety backup rather than relying on multiple identical devices.

Term [Citing Document(s)]	Definitions
Furnishings [STD 8719.11]	Consists of all movable articles, such as tables, chairs, desks, bookcases, draperies, cabinets, and decorations, required for use or as an ornament in a facility. • Interior Finish: Exposed material comprising walls, ceilings, wainscoting, and other interior building surfaces. It includes interior surfacing materials (such as paneling, carpeting, and wall coverings) applied directly to the walls, floors, and ceilings. Exposed insulating and acoustical materials are an interior finish. For purposes of controlling the hazards associated with combustible interior finish, the following classification system applies to Class A Materials having a Flame Spread Index not exceeding 25 and a Smoke Developed Index not exceeding 450, as determined by the test method described in NFPA 255. Carpets and rugs will also be Class A, if meeting the following criteria: • It has a value of CRF of 0.50 or above, as determined by the method described in NFPA 253. • It has a maximum specific optical density of not over 450 (flaming and non-flaming) as determined in NIST Technical Note 708 (Smoke Density Chamber). The critical specific optical density of 16 will not be reached in less than 30 seconds in both the flaming and non-flaming combustion. • Class B - Material having a Flame Spread Index between 26 and 75 and a Smoke Developed Index not exceeding 200, as determined by NFPA 255. Carpets and rugs will also be Class B if meeting the following criteria: • CRF between 0.25 and 0.50, as determined by the method described for Class A carpeting, and • Maximum specific optical density of not over 450, as described above. • Class C - Materials having a Flame Spread Index between 76 and 200 and a Smoke Developed Index not exceeding 450, as determined by NFPA 255. Carpets and rugs will also be Class C if they meet the following criteria. • Department of Commerce Standard for the Surface Flammability of Carpets and Rugs, FF 170, "Pill Test" • Maximum specific optical density of not over 450, as described above
fuse [STD 8719.24 Annex]	a system used to initiate an explosive train.
Gelling [STD 8739.1]	Formation of a semi-solid system consisting of a network of solid aggregates in which liquid is held; the initial gel-like solid phase that develops during the formation of a resin from a liquid.
general public [STD 8719.24 Annex]	all persons who are not in the launch-essential personnel or neighboring operations personnel categories; for a specific launch, the general public includes visitors, media, and other non-operations personnel at the launch site as well as persons located outside the boundaries of the launch site who are not associated with the specified launch; see also launch-essential personnel and neighboring operations personnel.
Geosynchronous Earth Orbit (GEO) [STD 8719.14]	A circular GEO with 0° inclination is a geostationary orbit about the Earth; i.e., the nadir point is fixed on the Earth's surface. The normal altitude of a circular GEO is 35,786 km and the inclination is normally +/- 15 degrees latitude.
Geosynchronous Transfer Orbit (GTO) [STD 8719.14]	A highly eccentric orbit with perigee normally within or near the LEO region altitude and apogee near or above GEO altitude.

Term [Citing Document(s)]	Definitions
GIDEP [NPR 8735.1]	This acronym stands for "Government-Industry Data Exchange Program." GIDEP is a cooperative information-sharing program between the U.S. Government, Canadian Government, and industry participants. The goal of GIDEP is to ensure that only reliable and conforming parts, materials, and software are in use on all Government programs and operations. GIDEP members share technical information essential to the research, design, development, production, and operational phases of the life cycle of systems, facilities, and equipment.
GIDEP Agency Action Notice [NPR 8735.1]	See the GIDEP Operations Manual for formal definition.
GIDEP Alert [NPR 8735.1]	See the GIDEP Operations Manual for formal definition.
GIDEP Notices [NPR 8735.1]	Term used within this NPR to collectively refer to GIDEP Alerts, GIDEP Safe-Alerts, GIDEP Problem Advisories, and GIDEP Agency Action Notices.
GIDEP Problem Advisory [NPR 8735.1]	See the GIDEP Operations Manual for formal definition.
GIDEP Representative [NPR 8735.1]	See the GIDEP Operations Manual for formal definition.
GIDEP Safe-Alert [NPR 8735.1]	See the GIDEP Operations Manual for formal definition.
Glass Transition [STD 8739.1]	A reversible change in an amorphous polymer or in amorphous regions of a partially crystalline polymer from (or to) a viscous or rubbery condition to (or from) a hard and relatively brittle one. Not only do hardness and brittleness undergo rapid changes in this temperature region, but other properties, such as dissipation factor, thermal expansibility, and specific heat, also change rapidly.
Glass Transition Temperature (Tg) [STD 8739.1]	The approximate midpoint of the temperature range over which glass transition takes place. The observed transition temperature can vary significantly depending on the specific property chosen for observation and on details of the experimental technique (for example, rate of heating, frequency). Therefore, the observed Tg should be considered only an estimate.
Glass Transition Temperature (Tg) [STD 8739.5]	The temperature above which an amorphous polymer displays viscous behavior caused by chain slip.
Government Contract Quality Assurance [NPR 8735.2]	Quality assurance functions performed by, or for, the Government at the contract location to determine whether a contractor has fulfilled the contract obligations pertaining to contract quality. Safety, reliability, and maintainability functions are also included within the scope of this term.

Term [Citing Document(s)]	Definitions
Government Industry Data Exchange Program (GIDEP) [STD 8739.10]	An organization through which users and suppliers of products (EEE parts, mechanical parts, materials, software, etc.) and the government may exchange information, such as part design changes and failure experiences.
Government Mandatory Inspection Point (GMIP) [NPR 8735.2]	A specific step, sequence, or time in a product's life when a NASAmandated product assurance action (e.g., product examination, process witnessing, record review) must be performed by NASA, a delegated Government agency, or by a NASA quality assurance support contractor.
Grade [STD 8739.10]	A classification which designates EEE parts in terms of reliability , quality or screening level based on military or industry standards. Interchangeable with terms "level" and "class" when used in this context.
Graded Approach [NPR 8000.4]	A "graded approach" applies risk management processes at a level of detail and rigor that adds value without unnecessary expenditure of unit resources. The resources and depth of analysis are commensurate with the stakes and the complexity of the decision situations being addressed.
Grommet [STD 8739.4]	An insulator that covers sharp edges of holes through panels and partitions to protect wire insulation from cut-through damage.
Ground Operations Plan [NPR 8715.7]	A detailed description of the hazardous and safety critical operations associated with a payload (spacecraft) and its associated ground support equipment. It contains the payload project's ground processing information providing the basis by which payload safety approval is obtained from the PSWG and Range Safety, along with the Safety Data Package. A flow chart of operations (hazardous and non-hazardous) is usually included. The Ground Operations Plan may be a stand-alone document or part of the payload project's Safety Data Package.
Ground Support Equipment [NPR 8715.3] [NPR 8715.7]	Ground-based equipment used to store, transport, handle, test, check out, service, and control aircraft, launch vehicles, spacecraft, or payloads.
Ground Support Equipment [STD 8719.17]	Non-flight equipment, systems, or devices specifically designed and developed for a physical or direct functional interface with flight hardware and to which the requirements of NASA-STD-5005 may apply.
Ground Support Equipment [NPR 8735.1]	Ground-based equipment used to store, transport, handle, test, check out, service, or control aircraft, launch vehicles, spacecraft, or payloads (NASA-STD-8709.22).
Ground Support Equipment (GSE) [STD 8739.10]	Non-flight equipment, systems, or devices specifically designed and developed for a direct physical or functional interface with flight hardware.

Term [Citing Document(s)]	Definitions
Ground Systems (GS) [NPR 8735.1]	This line of classification collectively refers to the ground-based hardware, software, development processes, and ground operations associated with supporting a vehicle or instrument operating in space. Implementation practices differ significantly from the flight hardware but are kept commensurate with the overall risk posture for the mission. Note that this category is distinct from Ground Support Equipment, which is equipment that is used to interface with flight hardware while it is on the ground. (examples: GOES-R GS, JPSS GS, SGSS)
Ground-based PVS [STD 8719.17]	All PVS, including Ground Support Equipment (GSE) PVS, and PVS based on barges, ships, or other transport vehicles, not specifically excluded in paragraph 4.2 of this standard. Flight PVS used for their intended purpose aboard active air or space craft, even though on the ground, are not included in this definition, but flight PVS converted to ground use are included.
Grounding [STD 8719.12]	Electrical. In the context of this document, electrical grounding refers to connections between conductive materials and structure to the earth for protection from transients caused by ESD or lightning.
handling structures [STD 8719.24 Annex]	structures such as beams, plates, channels, angles, and rods assembled with bolts, pins, and/or welds; includes lifting, supporting and manipulating equipment such as lifting beams, support stands, spin tables, rotating devices, and fixed and portable launch support frames.
Hardware [STD 8729.1]	Items made of a material substance but excluding computer software and technical documentation.
hardware (computer) [STD 8719.24 Annex]	physical equipment used in processing; items made of a material substance but excluding computer software and technical documentation.
Harness [STD 8739.4]	One or more insulated wires or cables, with or without helical twist; with or without common covering, jacket, or braid; with or without breakouts; assembled with two or more electrical termination devices; and so arranged that as a unit it can be assembled and handled as one assembly.
Hazard [NPR 8621.1]	A state or a set of conditions, internal or external to a system, having the potential to cause harm.
Hazard [NPR 8705.2]	A state or a set of conditions, internal or external to a system, which has the potential to cause harm (Source - NPR 8715.3).
Hazard [NPR 8715.3]	A state or a set of conditions, internal or external to a system that has the potential to cause harm.
Hazard [STD 8719.25]	A state or condition that could potentially lead to an undesirable consequence (i.e., casualty or property damage).
Hazard [NPR 8715.7]	A state or a set of conditions, internal or external to a system, that has the potential to cause harm.
Hazard	Any condition that may result in the occurrence or contribute to the severity of an accident.

Term [Citing Document(s)]	Definitions
[STD 8719.12]	
Hazard [STD 8719.7]	Any real or potential condition that can cause injury or death, or damage to or loss of equipment or property.
Hazard [STD 8719.9]	Any real or potential condition that can cause injury or death to personnel or damage to or loss of equipment or property.
Hazard Analysis [NPR 8705.2]	The process of identifying hazards and their potential causal factors.
Hazard Analysis [NPR 8715.3] [NPR 8715.7]	Identification and evaluation of existing and potential hazards and the recommended mitigation for the hazard sources found.
Hazard Analysis [STD 8719.12]	Logical, systematic examination of an item, process, condition, facility, or system to identify and analyze the probability, causes, and consequences of potential or real hazards.
Hazard Analysis [STD 8719.13]	[1] Identification and evaluation of existing and potential hazards and the recommended mitigation for the hazard sources found. [2] The process of identifying hazards and their potential causal factors.
hazard analysis [STD 8719.24 Annex]	the identification and evaluation of existing and potential hazards and the recommended mitigation for the hazard sources found; the process of identifying hazards and their potential casual factors.
Hazard Area [STD 8719.25]	A defined region of land, water, or airspace within which hazards exist or have the potential to exist during a range flight operation such that the risks associated with the hazards may be mitigated by controlling access to the defined region.
hazard area [STD 8719.24 Annex]	an area where known products can cause harm to the on- and off-base public.
Hazard Cause [STD 8719.7]	Any item that creates or significantly contributes to the existence of a hazard.
Hazard Control [NPR 8715.3] [NPR 8715.7]	Means of reducing the risk of exposure to a hazard.
Hazard Effects [STD 8719.7]	The potential detrimental consequences of the hazard.
hazard proof [STD 8719.24 Annex]	a method of making electrical equipment safe for use in hazardous locations; these methods include explosion proofing, intrinsically safe, purged, pressurized, and non-incendive and must be rated for the degree of hazard present.
Hazard Report [NPR 8715.7]	Hazard reports are an efficient means of summarizing for each identified hazard, the ways by which it can be caused, what controls are in place to prevent each cause, and the methods used to verify the performance of the hazard controls and compliance with associated safety design requirements. A hazard report is often used to document the results of a hazard analysis.

Term [Citing Document(s)]	Definitions
hazard severity [STD 8719.24 Annex]	the categorization of severity based on potential consequences and probabilities.
hazard, hazardous [STD 8719.24 Annex]	equipment, system, operation, or condition with an existing or potential condition that may result in a mishap; a state or a set of conditions, internal or external to a system, that has the potential to cause harm.
hazardous facility or structure [STD 8719.24 Annex]	a facility or structure used to store, handle, or process hazardous materials or systems and/or perform hazardous operations.
Hazardous Fragment [STD 8719.12]	A hazardous fragment is one having an impact energy of 58 ft-lb or greater.
Hazardous Fragment Distance (HFD) [STD 8719.12]	The distance at which the density of hazardous fragments becomes 1 per 600 ft2 (56 m2)
hazardous leak before burst [STD 8719.24 Annex]	a pressure vessel that exhibits a leak before burst failure mode and contains a hazardous material.
Hazardous Material [NPR 8715.3]	Defined by law as "a substance or materials in a quantity and form which may pose an unreasonable risk to health and safety or property when transported in commerce" (49 U.S.C S 5102). The Secretary of Transportation has developed a list of materials that are hazardous which may be found in 49 CFR pt. 172.101. Typical hazardous materials are those that may be highly reactive, poisonous, explosive, flammable, combustible, corrosive, radioactive, produce contamination or pollution of the environment, or cause adverse health effects or unsafe conditions.
Hazardous Material [NPR 8715.7]	Defined by law as "a substance or materials in a quantity and form which may pose an unreasonable risk to health and safety or property when transported in commerce" (49 U.S.C § 5102, Transportation of Hazardous Materials; Definitions). The Secretary of Transportation has developed a list of materials that are hazardous which may be found in 49 CFR Part 172.101. Typical hazardous materials are those that may be highly reactive, poisonous, explosive, flammable, combustible, corrosive, radioactive, produce contamination or pollution of the environment, or cause adverse health effects or unsafe conditions.
Hazardous Material [STD 8719.12]	Any material, item, or agent (biological, chemical, radiological, and/or physical), which has the potential to cause harm to humans, animals, or the environment, either by itself or through interaction with other elements or factors.
hazardous materials [STD 8719.24 Annex]	defined by law as "a substance or materials in a quantity and form which may pose an unreasonable risk to health and safety or property when transported in commerce" (49 U.S.C S 5102, Transportation of Hazardous Materials; Definitions). The Secretary of Transportation has developed a list of materials that are hazardous which may be found in 49 CFR Part 172.101. Typical hazardous materials are those that may be highly reactive, poisonous, explosive, flammable, combustible, corrosive, radioactive, produce contamination or pollution of the environment, or cause adverse health effects or unsafe conditions.

Term [Citing Document(s)]	Definitions
Hazardous Operation [NPR 8715.7]	Any operation involving material or equipment that has a high potential to result in loss of life, serious injury to personnel, or damage to systems, equipment, or facilities.
Hazardous Operation Safety Certification [NPR 8715.3]	Certification required for personnel who perform those tasks that potentially have an immediate danger to the individual (death/injury) if not done correctly, could create a danger to other individuals in the immediate area (death or injury), and present a danger to the environment.
Hazardous Operation/Work Activity [NPR 8715.3]	Hazardous Operation/Work Activity. Any operation or other work activity that, without implementation of proper mitigations, has a high potential to result in loss of life, serious injury to personnel or public, or damage to property due to the material or equipment involved or the nature of the operation/activity itself. .
hazardous operations [STD 8719.24 Annex]	those operations classified as hazardous according to the following criteria: (1) consideration of the potential or kinetic energy involved; (2) changes such as pressure, temperature, and oxygen content in ambient environmental conditions; (3) presence of hazardous materials; for example, operations involving equipment or systems with potential for a release of energy or hazardous material that can result in a mishap.
Hazardous Operations Support [STD 8719.24 Annex]	a Western Range contractor responsible for specific security operations.
hazardous pressure systems [STD 8719.24 Annex]	the systems used to store and transfer hazardous fluids such as cryogens, flammables, combustibles, hypergols; systems with operating pressures that exceed 250 psig; systems with stored energy levels exceeding 14,240 ft lb; systems that are identified by Range Safety as safety critical; see also safety critical.
hazardous procedure [STD 8719.24 Annex]	a designation for a particular type of Range User procedure; a document containing specific steps in sequential order used to safely process hazardous materials or conduct hazardous operations; hazardous procedures have specific content requirements delineated in Volume 6, Attachment 2 and require Range Safety approval.
Hazardous Waste [STD 8719.12]	Any hazardous material that is discarded; disposed of; burned or incinerated; accumulated; stored; or used in a manner constituting disposal.
Heritage Hardware [STD 8739.10]	Hardware whose design has been previously qualified and used in space applications, and was accepted for use by a NASA program or project.
High Density Traffic [STD 8719.12]	Traffic routes having 10,000 or more car and/or rail passengers per day, or 2,000 or more ship passengers per day.
High Explosive (HE) [STD 8719.12]	An explosive (as denoted by its Class and Division; e.g., 1.1 through 1.6) in which the transformation from its original composition and form, once initiated, proceeds with virtually instantaneous and continuous speed through the total mass, accompanied by rapid evolution of a large volume of gas and heat, causing very high pressure and widespread shattering effect.
High Performance Magazine (HPM) [STD 8719.12]	An earth-berm, 2-story magazine with internal non-propagation walls designed to reduce the maximum credible event (MCE).

Term [Citing Document(s)]	Definitions
High Voltage [STD 8739.1]	An application voltage that will support coronal discharge or the development of charged plasma due to the environment's atmospheric conditions (i.e. elemental gasses or vacuum, and pressure) unless mitigations such as rounded surfaces at interconnections and insulative layers of staking or conformal coating are applied.
high voltage exploding bridgewire [STD 8719.24 Annex]	an initiator in which the bridgewire is designed to be exploded (disintegrated) by a high energy electrical discharge that causes the explosive charge to be initiated.
High-Visibility Mishap or Close Call [NPR 8621.1]	A mishap or close call, regardless of the amount of property damage or personnel injury, that the Administrator; Chief, Safety and Mission Assurance, Office of Safety and Mission Assurance; Center Director, Executive Director, Office of Headquarters Operations; Aircraft Management Division Director; or Center Safety and Mission Assurance Director judges to possess a high degree of safety risk, programmatic impact or public, media, or political interest including, but not limited to, mishaps and close calls affecting flight hardware or software, or completion of critical mission milestones.
Hoist [STD 8719.9]	A machinery unit device used for lifting and lowering a load.
hoist angle [STD 8719.24 Annex]	an angle at which the load line is pulled during hoisting.
Hoist-Supported Personnel Lifting Device [STD 8719.9]	Device specifically designed to lift and lower persons via a hoist. These devices include hoist-supported platforms where personnel occupy the platform during movement. These devices do not include elevators, lifting personnel with a crane, mobile aerial platform, or platforms hoisted unoccupied to a position and anchored or restrained to a stationary structure before personnel occupy the platform (refer to Personnel Access Platform).
hold [STD 8719.24 Annex]	a temporary delay in the countdown, test, or practice sequence for any reason.
holdfire [STD 8719.24 Annex]	an interruption of the ignition circuit of a launch vehicle.
Holding Brake [STD 8719.9]	A brake that automatically prevents motion when power is off.
Holding Yard [STD 8719.12]	A holding area for rail cars, trucks, or trailers used for temporary storage of vehicles containing explosives and other dangerous materials prior to shipment or transfer to a more permanent storage area.
Hot Work [STD 8719.12]	Any operation requiring the use of a flame-producing device, an electrically heated tool, or a mechanical tool that can produce sparks or heat thereby providing an initiation stimulus.
Hull Loss [NPR 8621.1]	An aircraft damaged to the extent that repair is not economically feasible. This includes destroyed and missing aircraft (exception: unmanned aircraft).

Term [Citing Document(s)]	Definitions
Human Error [NPR 8621.1]	An action unintended or undesired by the human or a failure on the part of the human to perform a prescribed action within specified limits of accuracy, sequence, or time that fails to produce the expected result and has led or has the potential to lead to an undesired outcome.
Human Error [NPR 8705.2]	Either an action that is not intended or desired by the human or a failure on the part of the human to perform a prescribed action within specified limits of accuracy, sequence, or time that fails to produce the expected result and has led or has the potential to lead to an unwanted consequence.
Human Error Analysis (HEA) [NPR 8705.2]	A systematic approach to evaluate human actions, identify potential human error, model human performance, and qualitatively characterize how human error affects a system. HEA provides an evaluation of human actions and error in an effort to generate system improvements that reduce the frequency of error and minimize the negative effects on the system. HEA is the first step in Human Risk Assessment and is often referred to as qualitative Human Risk Assessment.
Human Factors [NPR 8621.1]	a. A body of scientific facts about human characteristics, capabilities, and behavior. The term includes, but is not limited to, principles and applications in the areas of human engineering, personnel selection, training, life support, job performance aids, and human performance evaluation. b. A body of information about human abilities, human limitations, and other human characteristics from a physical and psychological perspective relevant to the design, operations, and maintenance of complex systems.
Human Factors Analysis [NPR 8621.1]	The study of how people interact with their environment. Physiological, psychological, and organizational behaviors are evaluated. Human factors analysis is an important component of mishap investigation. Determining why, how, and where human behaviors contributed to mishaps and close calls is key to preventing future mishaps.
Human Factors Investigator [NPR 8621.1]	An investigator with expertise in human factors engineering and mishap causation who has primary responsibility to assist in data collection and analysis; determine the manner in which human factors caused or contributed to the mishap or close call; evaluate relevant human error and determine its root causes; and generate recommendations to eliminate or reduce error occurrence or minimize the error's negative effects to prevent the occurrence of a similar mishap.
Human Health Management and Care [NPR 8705.2]	The set of activities, procedures, and systems that provide (1) environmental monitoring and human health assessment; (2) health maintenance and countermeasures; and (3) medical intervention for the diagnosis and treatment of injury and illness.
Human Performance [NPR 8705.2]	The physical and mental activity required of the crew and other participants to accomplish mission goals. This includes the interaction with equipment, computers, procedures, training material, the environment, and other humans.
Human-Rated Space System [NPR 8705.2]	A human-rated system accommodates human needs, effectively utilizes human capabilities, controls hazards with sufficient certainty to be considered safe for human operations, and provides the capability to safely recover from emergency situations. The concept of human-rating a space system entails three fundamental tenets: 1. Human-rating is the process of evaluating and assuring that the total system can safely conduct the required human missions.

Term [Citing Document(s)]	Definitions
	2. Human-rating includes the incorporation of design features and capabilities that accommodate human interaction with the system to enhance overall safety and mission success.
	3. Human-rating includes the incorporation of design features and capabilities to enable safe recovery of the crew from hazardous situations.
Human-Rating Certification [NPR 8705.2]	Human-Rating Certification is the documented authorization granted by the NASA Administrator that allows the program manager to operate the space system within its prescribed parameters for its defined reference missions. Human-Rating Certification is obtained prior to the first crewed flight (for flight vehicles) or operational use (for other systems).
Human-Rating Certification Package [NPR 8705.2]	See Appendix D.
Human-Rating Process [NPR 8705.2]	The process steps used to achieve a human-rated space system. These steps include human safety risk identification, reduction, control, visibility, and program management acceptance criteria. Acceptable methods to assess the risk to human safety include qualitative and/or quantitative methods such as hazards analysis, fault tree analysis, human error analysis, probabilistic risk assessment, and failure modes and effects analysis.
Human-System Integration [NPR 8705.2]	The process of integrating human operations into the system design through analysis, testing, and modeling of human performance, interface controls/displays, and human-automation interaction to improve safety, efficiency, and mission success.
Hydraset [STD 8719.24 Annex]	the trade name for a closed circuit hydraulically operated instrument installed between a crane hook and load that allows precise control of lifting operations and provides an indication of applied load; precision load positioning device.
hydraulic [STD 8719.24 Annex]	operated by water or any other liquid under pressure; includes all hazardous fluids as well as typical hydraulic fluids that are normally petroleum-based.
Hydraulics [STD 8719.17]	Hydraulic systems using commercially available hydraulic fluid.
hydrogen embrittlement [STD 8719.24 Annex]	a mechanical-environmental failure process that results from the initial presence or absorption of excessive amounts of hydrogen in metals, usually in combination with residual or applied tensile stresses.
hygroscopic [STD 8719.24 Annex]	absorbs moisture from the air.
Hypergolic [STD 8719.12]	Self-igniting upon contact of fuel and oxidizer, without a spark or external aid.
hypergolic propellants [STD 8719.24 Annex]	Fluids that ignite spontaneously upon mixing for the purposes of propulsion and power, such as certain rocket fuels and oxidizers; Self-igniting upon contact of a fuel and an oxidizer, without a spark or external aid.

Term [Citing Document(s)]	Definitions
Idle Lifting Device [STD 8719.9]	Lifting device that has not been used for 12 months or more, or that has no projected use for the next 12 months.
igniter [STD 8719.24 Annex]	a device containing a specifically arranged charge of ready burning composition, usually black powder, used to amplify the initiation of a primer.
imminent danger [STD 8719.24 Annex]	any condition, operation, or situation that occurs on the range where a danger exists that could reasonably be expected to cause death or serious physical harm, immediately or before the imminence of such danger can be eliminated through control procedures; these situations also include health hazards where it is reasonably expected that exposure to a toxic substance or other hazard will occur that will cause harm to such a degree as to shorten life or cause a substantial reduction in physical or mental efficiency even though the resulting harm may not manifest itself immediately.
Impact [NPR 8735.1]	The issue presented in the NASA Advisory, GIDEP Notice, or other released document potentially has an adverse effect on operation of the system being evaluated.
Incident [NPR 8621.1]	An occurrence of a mishap or close call.
Incident Commander [NPR 8621.1]	The person responsible for directing or controlling resources by means of explicit legal, Agency, or delegated authority. The incident commander is responsible for all aspects of incident response including developing objectives, managing operations, setting priorities, and defining the Incident Command System organization for the particular response.
Inclination [STD 8719.14]	The angle an orbital plane makes with the Earth's equatorial plane.
independent [STD 8719.24 Annex]	not capable of being influenced by other systems.
Independent Verification and Validation [NPR 8715.3]	Test and evaluation process by an independent third party.
Independent Verification and Validation (IV&V) [STD 8739.8]	Verification and validation performed by an organization that is technically, managerially, and financially independent. IV&V, as a part of software assurance, plays a role in the overall NASA software risk mitigation strategy applied throughout the life cycle, to improve the safety and quality of software.
indication [STD 8719.24 Annex]	the response or evidence from the application of a nondestructive examination including visual inspection.
Individual Risk [STD 8719.25]	The probability of an individual from a certain group (or subgroup) at a specific location suffering a casualty from exposure to hazards from a given event during an established period (e.g., a launch). Individual risk is stated as a Probability of Casualty (Pc).
Inert [STD 8719.12]	Containing no explosive or chemical agents. Inert material shall show no incompatibility with energetic material with which it may be combined when tested by recognized compatibility tests.

Term [Citing Document(s)]	Definitions
Inhabited Building Distance (IBD) [STD 8719.12]	Minimum allowable distance between an inhabited building and an explosive facility. IBDs are used between explosives facilities and administrative areas, operating lines with dissimilar hazards, explosive locations and other exposures, and explosive facilities and Center boundaries, and define the restricted zone into which non-essential personnel may not enter.
Inhabited Buildings [STD 8719.12]	A building or structure occupied in whole or in part by human beings, or where people are accustomed to assemble, both within and outside of Government establishments. Land outside the boundaries or local restrictive easement estate of NASA establishments is considered as inhabited buildings.
Inhibit [NPR 8715.3]	Design feature that prevents operation of a function.
Inhibit [NPR 8715.7]	An independent and verifiable mechanical and/or electrical device that prevents a hazardous event from occurring; the device has direct control and is not the monitor of such a device.
inhibit [STD 8719.24 Annex]	an independent and verifiable mechanical or electrical device that prevents a hazardous event from occurring; device has direct control and is not the monitor of such a device. An inhibit is a physical interruption or barrier between an energy source and the action or function that would take place if the energy source is released. Examples would be a relay or transistor between a battery and a pyrotechnic initiator, or a latch valve between a pressurized propellant tank and a thruster. Note: An inhibit control is a device or function that operates an inhibit. Controls do not satisfy the inhibit or failure tolerance requirements for hazardous functions. An example, as stated in Volume 3 paragraph 3.2.8 of this document, is software. Software is considered an inhibit operator control, not an inhibit.
initial flaw [STD 8719.24 Annex]	a flaw in a structural material before the application of load and/or environment.
Initiating Event [NPR 8621.1]	An active energy transfer event from a hazard with the potential to affect a valued target and lead to an undesired outcome.
initiator [STD 8719.24 Annex]	includes low voltage electroexplosive devices and high voltage exploding bridgewire devices.
Insertion Loss [STD 8739.5]	The optical attenuation caused by the insertion of an extra optical component into an optical system.
Insertion Tool [STD 8739.4]	A device used to install contacts into a contact cavity in a connector insert.
Inservice Inspections [STD 8719.17]	Those inspections, examinations, or tests specified in the inspection plan as determined in this document.
Insight [STD 8739.8]	Surveillance mode requiring the monitoring of acquirer-identified metrics and contracted milestones. Insight is a continuum that can range from low intensity, such as reviewing quarterly reports, to high intensity, such as performing surveys and reviews.

Term [Citing Document(s)]	Definitions
Inspection [STD 8739.9]	A technical evaluation process during which a product is examined with the purpose of finding and removing defects and/or discrepancies as early as possible in the software life cycle.
Inspection Package [STD 8739.9]	The physical and/or electronic collection of software products and corresponding documentation presented for inspection as well as required and appropriate reference materials.
Inspection Report [STD 8739.9]	A report used to document and communicate the status (such as time and defect data) of a software formal inspection.
Inspector [STD 8739.9]	Participant in an inspection.
Installation Load, maximum [STD 8739.5]	The maximum load which can be applied along the axis of a cable during installation without breaking fibers or causing a permanent increase in the cable attenuation.
Institutional Risks [NPR 8000.4]	Risks to infrastructure, information technology, resources, personnel, assets, processes, operations, occupational safety and health, environmental management, security, or programmatic constraints that affect capabilities and resources necessary for mission success, including institutional flexibility to respond to changing mission needs and compliance with internal (e.g., NASA) and external requirements (e.g., Environmental Protection Agency or Occupational Safety and Health Administration regulations).
Institutional, Facility, Operational Safety Audit [NPR 8705.6]	An independent audit of NASA Center compliance with institutional, facility, and operational SMA requirements.
Interchange Yard [STD 8719.12]	A location set aside for exchange of rail cars or trailers between a common carrier and NASA.
Interface [STD 8739.9]	The boundary, often conceptual, between two or more functions, systems, or items, or between a system and a facility, at which interface requirements are set.
Interfacial Seal [STD 8739.4]	A sealing of mated connectors over the whole area of the interface to provide sealing around each contact.
Interferometer [STD 8739.5]	An instrument that employs the interference of light waves for purposes of measurement.

Term [Citing Document(s)]	Definitions
Interim Response Team [NPR 8621.1]	A team called to the mishap scene immediately after an incident to secure the scene; document the scene using photography, video, sketches, and debris mapping; identify witnesses; collect written witness statements and contact information; preserve evidence; impound evidence at the scene and other NASA locations as needed; collect debris; implement the chain-of-custody process for the personal effects of the injured and deceased; notify the Public Affairs Office about casualties, damages, and potential hazards to the public and NASA personnel; advise the supervisor if drug testing should be initiated; and provide all information and evidence to the investigating authority. The team is considered interim because it operates as a short-term response team and concludes its mishap response activities when the official NASA-appointed investigating authority takes control.
Interlock [NPR 8715.3]	Hardware or software function that prevents succeeding operations when specific conditions are satisfied.
Intermagazine Distance (IMD) [STD 8719.12]	Distance to be maintained between two explosives storage locations.
Intermediate Cause [NPR 8621.1]	An event or condition that existed before the proximate cause, directly resulted in its occurrence, and if eliminated or modified, would have prevented the proximate cause from occurring.
interrupter [STD 8719.24 Annex]	a mechanical barrier in a fuse that prevents transmission of an explosive effect to some elements beyond the interrupter.
Intraline Distance (ILD) [STD 8719.12]	The minimum distance allowed between any two operating locations or other designated exposures. This distance is expected to prevent propagation.
intrinsically safe [STD 8719.24 Annex]	incapable of producing sufficient energy to ignite an explosive atmosphere and two fault tolerant against failure with single fault tolerance against its most hazardous failure at 1.5 times the maximum voltage or energy.
Investigating Authority [NPR 8621.1]	The individual mishap investigator, mishap investigation team, or mishap investigation board authorized to conduct an investigation for NASA. This includes the mishap investigation board chairperson, voting members, and ex officio, but does not include the advisors and consultants.
ionizing radiation [STD 8719.24 Annex]	gamma and X-rays, alpha and beta particles and neutrons.
Jack [STD 8719.9]	A mechanism with a base and load point designed for controlled linear movement.
Jacket [STD 8739.4]	The outermost layer of insulating material of a cable or harness.
Jansky [STD 8719.14]	A unit of electromagnetic power density equal to 10-26 watts/m2/Hz.
Joint	A termination.

Term [Citing Document(s)]	Definitions
[STD 8739.4]	
Keep Out Areas [STD 8739.1]	Surfaces on PWAs that must remain free of polymeric material following bonding, staking, conformal coating or encapsulating processes to enable subsequent system assembly processes and system performance requirements (e.g. fastener or electrical ground interfaces). Masking is used to prevent polymeric material from coming into contact or covering keep out areas during polymeric applications.
Key Decision Point [NPR 8715.7]	(Per NPR 7120.5) An event where the Decision Authority (the Agency's responsible individual who authorizes the transition of a program/project to the next life-cycle phase) determines the readiness of a program/project to progress to the next phase of the life cycle. As such, Key Decision Points serve as gates through which programs and projects must pass.
K-Factor [STD 8719.12]	K is a constant that is used to determine separation distance by the formula $d = KW1/3$, where W is the weight in pounds. The formula can be used to determine required distances between potential explosive sites (PESs) and exposed sites (ESs). This will normally appear as the letter "K" followed by a number, for example "K8," or "K30."
Knowledge Management [NPR 8000.4]	Knowledge management is getting the right information to the right people at the right time and helping people create knowledge and share and act upon information in ways that will measurably improve the performance of NASA and its partners.
Laboratory Operations [STD 8719.12]	Experimental study, testing, and analysis of small quantities of energetic materials. Manufacturing processes with small quantities of materials are not included. Includes operations in a laboratory where the total quantity of 1.1 Class/Division explosive materials in the room does not exceed 200 grams. For maximum quantities of other Class 1 Divisions, use TNT equivalencies where the comparable quantity for the Class 1 Divisions is determined by the TNT equivalency.
Landing [NPR 8705.2]	The final phase or region of flight to Earth/Lunar surface consisting of transition from descent, to an approach, touchdown, and coming to rest.
Landing Site [NPR 8715.5]	The earth location on which a vehicle impacts, lands, or is recovered.
Landing Site [STD 8719.25]	The location on which a vehicle impacts, lands, or is recovered.
Laser [STD 8739.5]	A device that produces coherent optical radiation by stimulated emission and amplification.
Launch [NPR 8621.1]	To place a vehicle and any payload from Earth in a suborbital trajectory, in Earth orbit, or in outer space.
Launch [NPR 8715.5]	To place a vehicle, payload, or astronauts from Earth in a suborbital trajectory, in Earth orbit or in outer space. For an orbital mission, launch begins with lift-off and ends with orbital insertion. For a suborbital mission, launch begins with lift-off and ends with landing/final impact of all vehicle components.

Term [Citing Document(s)]	Definitions
Launch [STD 8719.25]	To place a vehicle and any payload from Earth in an altitude (balloons or UAVs), in a suborbital trajectory, in Earth orbit, or in outer space. For an orbital mission, launch begins with lift-off and ends with orbital insertion. For a suborbital mission, launch begins with lift-off and ends with landing/final impact of all vehicle components.
launch abort [STD 8719.24 Annex]	the termination of a launch sequence in an unplanned manner or the failure of the launch vehicle to liftoff for reasons not immediately known.
launch area [STD 8719.24 Annex]	the facility or location where launch vehicles and payloads are processed and launched; includes any supporting sites; also known as launch head. The launch area extends to the over-water areas used during submarinelaunched ballistic missile intercontinental ballistic missile tests and launches where the range controls the launch for countdown.
launch area safety [STD 8719.24 Annex]	safety requirements involving risks limited to personnel and/or property located on the launch base; involves multiple commercial users, government tenants, or United State Air Force squadron commanders; this is the on-base component of public safety.
launch complex [STD 8719.24 Annex]	a defined area that supports launch vehicle or payload operations or storage; includes launch pads and/or associated facilities.
launch complex safety [STD 8719.24 Annex]	safety requirements involving risk that is limited to personnel and/or property located within the well-defined confines of a launch complex, facility, or group of facilities; for example, within the fence line; involves risk only to those personnel and/or property under the control of the control authority for the launch complex, facility, or group of facilities.
launch processing [STD 8719.24 Annex]	all preflight preparation of a launch vehicle at a launch site, including buildup of the launch vehicle, integration of the payload, and fueling.
Launch Site [NPR 8715.5] [STD 8719.25]	The location from which a launch takes place. This includes land, air, or a sea-based position.
launch site [STD 8719.24 Annex]	the specific geographical location from which a launch takes place.
Launch Vehicle [STD 8719.14]	Any space transportation mode, including expendable launch vehicles (ELVs), reusable launch vehicles (RLVs), and the Space Shuttle.
launch vehicle [STD 8719.24 Annex]	a vehicle that carries and/or delivers a payload to a desired location; a generic term that applies to all vehicles that may be launched from the Eastern and Western ranges, including but not limited to airplanes; all types of space launch vehicles; manned space vehicles; missiles; rockets and their stages; probes, aerostats, and balloons; drones; remotely piloted vehicles; projectiles, torpedoes, and air-dropped bodies.
LDEM-Lifting Device and Equipment Manager [STD 8719.24 Annex]	NASA person responsible for overall management of the installation lifting devices and equipment program, coordinating with appropriate personnel at their installation on lifting issues and providing their installation's position on lifting devices and equipment safety issues.
lead angle [STD 8719.24 Annex]	an angle in which the load line is pulled during hoisting. Commonly used to refer to an angle in line with the grooves in the drum or sheaves.

Term [Citing Document(s)]	Definitions
lead time [STD 8719.24 Annex]	the time between the beginning of a process or project and the appearance of its results.
leak before burst [STD 8719.24 Annex]	a failure mode in which it can be shown that any initial flaw will grow through the wall of a pressure vessel or pressurized structure and cause leakage rather than brittle fracture/burst before leak; normally determined at or below maximum expected operating pressure.
Lessons Learned [NPR 8621.1]	The written description of knowledge or understanding gained by experience, whether positive such as a successful test or mission, or negative such as a mishap or failure.
Level A Instructor [STD 8739.6]	Instructor who teaches one or more of NASA-STD-8739.1, NASA-STD-8739.2, NASA-STD-8739.3, NASA-STD-8739.4, or NASA-STD-8739.5 courses to operators, inspectors, and Level B instructors (See A.2.1.g). The local ESD Control Plan may choose to define and use a NASA Level A Instructor classification in its training section.
Level B Instructor [STD 8739.6]	Instructor who teaches one or more of NASA-STD-8739.1, NASA-STD-8739.2, NASA-STD-8739.3, NASA-STD-8739.4, or NASA-STD-8739.5 courses to operators and inspectors (see A.2.1.d.). The local ESD Control Plan may choose to define and use a Level B Instructor classification in its training section.
Level of Repair Analysis (LORA) [STD 8729.1]	An analytical methodology used to assist in developing maintenance concepts and establishing the maintenance level at which components will be replaced, repaired, or discarded based on economic/noneconomic constraints and operational readiness requirements. Also known as an Optimum Repair Level Analysis (ORLA).
License [STD 8719.12]	Formal documented permission from the ESO to operate a Licensed Explosive Location.
Licensed Explosive Locations [STD 8719.12]	Locally licensed locations within NASA's control where explosives are used or stored for use.
Licensed Operator [STD 8719.9]	See Licensed Personnel.
Licensed Personnel [STD 8719.9]	Individuals documented by the LDEM as meeting the personnel licensing requirements of this standard. Licensed personnel may be referred to as certified personnel or certified operators in other regulations and VCS.
Life Cycle [NPR 8705.2]	The totality of a program or project extending from formulation through implementation encompassing the elements of design, development, verification, production, operation, maintenance, support and disposal.
Life-Threatening Injury [NPR 8621.1]	An injury involving a substantial risk of death; loss or substantial functional impairment of a bodily member, organ, or mental faculty likely to be permanent; or an obvious disfigurement likely to be permanent.
Lifting [STD 8739.1]	Any separation of conformal coating from the PWA.

Term [Citing Document(s)]	Definitions
Lifting Device [STD 8719.9]	A generic term or modifier broadly used to refer to both equipment that actively lifts (cranes, powered industrial trucks, etc.) and individual pieces or assemblies of components used in the lifting process (slings, hoist-supported lifting devices, shackles, etc.).
Lifting Devices and Equipment (LDE) [STD 8719.9]	Devices, equipment, and their accessories used to lift, lower, and position a load.
Lifting Devices and Equipment Manager (LDEM) [STD 8719.9]	Person designated by the Center Director, responsible for managing the installation lifting devices and equipment program, coordinating with appropriate personnel at their installation on lifting issues, and providing their installation's position on lifting devices and equipment safety issues.
Likelihood [STD 8719.13]	Likelihood is the chance that something might happen. Likelihood can be defined, determined, or measured objectively or subjectively and can be expressed either qualitatively or quantitatively (using mathematics). [From ISO 31000 2009 Plain English Risk Management Dictionary.] For this document looking at the software contribution; likelihood does not solely represent a probability of the initiating software cause, as these are systematic faults; it is a qualitative estimate of the likelihood of the software fault to propagate to the hazard (top level event). Factors such as control autonomy are rolled into that likelihood.
Likelihood [NPR 8000.4]	Probability of occurrence.
limit load [STD 8719.24 Annex]	the calculated maximum loads to which a structure may be subjected during its lifetime of service; i.e., the applied load (static or dynamic) multiplied by applicable load amplification factors; see limit load (design load).
Line of Apsides [STD 8719.14]	The line connecting the apogee and perigee points in an orbit. This line passes through the center of the Earth.
Line of Nodes [STD 8719.14]	The line formed by the intersection of the orbit plane with the Earth's equatorial plane. This line passes through the center of the Earth. The ascending node is the point where a satellite crosses the equator from the southern hemisphere to the northern hemisphere.
lines [STD 8719.24 Annex]	the tubular pressure components of a pressurized system provided as a means for transferring fluids between components of the system.
Liquid Propellant [STD 8719.12]	Liquid and gaseous substances (fuels, oxidizers, or monopropellants) used for propulsion or operation of rockets and other related devices.
Listed Or Approved [STD 8719.11]	When referring to a material or device used in conjunction with fire protection, a product that has been tested by a recognized and independent research laboratory (e.g., Underwriters Laboratories and Factory Mutual), in accordance with generally accepted and standardized test methods and verified that it will perform adequately and dependably under adverse conditions.

Term [Citing Document(s)]	Definitions
Load [STD 8719.9]	The total weight of the items being supported, raised, or moved by a lifting device or equipment, including rigging hardware, slings, below-the-hook lifting devices, the load block for some mobile crane configurations, or any other attachments that are not taken into account when determining the rated capacity of the lifting device or equipment.
Load Brake [STD 8719.9]	A braking device that retards and controls the load during lowering and keeps the load from falling if the holding brake fails.
Load Measuring Device [STD 8719.9]	A device below the hook which is used to indicate the weight of the item being lifted.
Load Positioning Device [STD 8719.9]	Instrument installed between the hook and load to allow precise control of lifting operations (e.g., Hydra Sets®).
loading spectrum [STD 8719.24 Annex]	a representation of the accumulated loadings anticipated for the structure under all expected operating environments; significant transportation and handling loads are included.
Local Center [STD 8709.20]	The Center or component facility executing or hosting the applicable activity.
local safety authority [STD 8719.24 Annex]	approving organization designated and authorized to make safety decisions for a specific facility or launch site (i.e., Range Safety, LSP S&MA, PPF Safety, etc.).
Loosely Coupled Program/Project [STD 8709.20]	Programs/Projects that are not a part of Tightly Coupled Programs. (See NASA Memo 7120-81, paragraph 2.1.4, which updated NPR 7120.5D, for further definitions on program/project types.)
Lost Time Injury or Illness [NPR 8621.1]	A nonfatal traumatic injury resulting in any loss of time from work beyond the day or shift it occurred; or a nonfatal, non-traumatic illness or disease causing disability at any time.
Lot [STD 6008]	A collection of units or items (e.g., fasteners or inserts) manufactured from a homogeneous batch of material of the same continuous, uninterrupted production.
Lot Date Code (LDC) [STD 8739.10]	An identification code, usually marked on a EEE part and prescribed by the applicable specification, to identify parts which have been processed as a batch.
Low Earth Orbit (LEO) [STD 8719.14]	An orbit with a mean altitude less than or equal to 2000 km, or equivalently, an orbit with a period less than or equal to 127 minutes.
Low Explosives [STD 8719.12]	Propellants, which have a controlled rate of gas pressure, i.e. deflagration (subsonic).
Low Released Mass [STD 6008]	A fastener that meets the criteria specified in NASA-STD-5019, paragraph 4.1.1.1.
Low Traffic Density [STD 8719.12]	Traffic routes having less than 400 cars and/or rail passengers per day or less than 80 ship passengers per day.

Term [Citing Document(s)]	Definitions
Low-Risk Fracture Fastener (or Low-Risk Fastener) [STD 6008]	A fastener that meets the criteria specified in NASA-STD-5019, paragraph 4.1.1.12.
Magazine [STD 8719.12]	A structure designed or specifically designated for the storage of explosives.
Magazine Distance [STD 8719.12]	Minimum distance permitted between any two storage magazines. The distance required is determined by the type(s) of magazine and also the type and quantity of explosives stored therein.
Maintainability [STD 8729.1]	A measure of the ease and rapidity with which a system or equipment can be restored to operational status. It is characteristic of equipment design and installation, personnel availability in the required skill levels, adequacy of maintenance procedures and test equipment, and the physical environment under which maintenance is performed. One expression of maintainability is the probability that an item will be retained in or restored to a specified condition within a given period of time, when the maintenance is performed in accordance with prescribed procedures and resources.
Maintenance [STD 8729.1]	All actions necessary for retaining an item in, or restoring it to, a specified condition.
Maintenance Analysis [STD 8729.1]	The process of identifying required maintenance functions by analysis of the design, and to determine the most effective means to accomplish those functions.
major leak or spill [STD 8719.24 Annex]	a leak or spill that could affect regions beyond the immediate work area, constitute a hazard to personnel, or involve damage to facilities or equipment; a major leak or spill is more than one gallon.
major mishap [STD 8719.24 Annex]	an event or incident that has the potential of resulting in a fatality or major damage such as the loss of a processing facility, launch complex, launch vehicle, or payload.
Major Nonconformance [NPR 8735.1]	A nonconformance, other than critical, that is likely to result in failure of the supplies or services, or to materially reduce the usability of the supplies or services for their intended purpose.
mandatory (in reference to instrumentation or capability) [STD 8719.24 Annex]	a system that must be made operationally ready to support Range Safety and be fully mission capable before entering the plus count.
Mandatory SMA Standard [STD 8709.20]	A standard (NASA owned/developed or otherwise) that is directed to be used by a requirement in an NPD, NPR, or other mandatory SMA standard.
Manual Control [NPR 8705.2]	The crew's ability to bypass automation in order to exert direct control over a space system or operation. For control of a spacecraft's flight path, manual control is the ability for the crew to effect any flight path within the capability of the flight control system. Similarly, for control of a spacecraft's attitude, manual control is the ability for the crew to effect any attitude within the capability of the flight/attitude control system.

Term [Citing Document(s)]	Definitions
Manufacturer's Test Report (MTR) [STD 6008]	A document that is produced by the fastener manufacturer that certifies information required by the applicable fastener specification.
Manufacturing Documentation [STD 8739.1] [STD 8739.6]	Instructions, drawings, specifications, work orders, travelers and all other documents provided to manufacturing operators and inspectors defining the intended design, manufacturing methods, and quality controls.
Margin of Safety [NPR 8715.3]	Deviation of the actual (operating) factor of safety from the specified factor of safety. Can be expressed as a magnitude or percentage relative to the specified factor of safety.
margin of safety [STD 8719.24 Annex]	the percentage by which the allowable load (stress) exceeds the limit load (stress) for specific design conditions; Yield Margin of Safety = [(Yield Strength/Limit Load Stress) x (Yield Factor of Safety)] - 1; Ultimate Margin of Safety = [(Ultimate Strength/Limit Load Strength) x (Ultimate Factor of Safety)] – 1.
marginal hazard [STD 8719.24 Annex]	a hazard, condition or event that may cause minor injury or minor occupational illness to personnel.
Mass Detonation/Explosion [STD 8719.12]	Virtually instantaneous explosion of a mass of explosives when only a small portion is subjected to fire, severe concussion or impact, the impulse of an initiating agent, or to the effect of a considerable discharge of energy from an outside stimulus. Also refers to the instantaneous propagation of an explosion between multiple explosives items such that blast overpressure effects are combined into a single enhanced blast wave.
Mate [STD 8739.4]	The joining of two connectors.
materials, brittle [STD 8719.24 Annex]	those materials that undergo little plastic tensile or shearing deformation before rupture; see also ductile behavior.
materials, ductile [STD 8719.24 Annex]	those materials that undergo considerable plastic tensile or shearing deformation before rupture, and have sufficient notch toughness to fracture in a ductile manner at operating temperatures and under impact loading; see ductile behavior in this volume and Mechanics of Materials in References.
maximum allowable working pressure [STD 8719.24 Annex]	the maximum pressure at which a component or system can continuously operate based on allowable stress values and functional capabilities.
Maximum Credible Event (MCE) [STD 8719.12]	In hazards evaluation, the MCE from a hypothesized accidental explosion, fire, or agent release is the worst single event that is likely to occur from a given quantity and disposition of explosives, chemical agents, or reactive material. The event must be realistic with a reasonable probability of occurrence considering the explosion propagation, burning rate characteristics, and physical protection given to the items involved. The MCE evaluation on this basis may then be used as a basis for effects calculations and casualty prediction.

Term [Citing Document(s)]	Definitions
maximum expected operating pressure [STD 8719.24 Annex]	the highest pressure that a pressure vessel, pressurized structure, or pressure component is expected to experience during its service life and retain its functionality, in association with its applicable operating environments; synonymous with maximum operating pressure or maximum design pressure includes the effect of temperature, pressure transients and oscillations, vehicle quasi-steady, and dynamic accelerations and relief valve operating variability.
May [STD 8709.20]	Good practices, guidance, or options are specified with the nonemphatic verbs "should," "may," or "can (from NASA-STD 0005).
Means of Egress [STD 8719.11]	A means of egress is a continuous and unobstructed way of travel from any point in a building or structure to a public way. A means of egress comprises the vertical and horizontal travel and includes intervening room spaces, doorways, hallways, corridors, passageways, balconies, ramps, stairs, enclosures, lobbies, escalators, horizontal exits, courts, and yards.
Measuring and Test Equipment [STD 8739.12]	The measuring instrument, measurement standard, reference material, or auxiliary apparatus, or a combination thereof, necessary to realize a measurement process.
Medium Traffic Density [STD 8719.12]	Traffic routes having 400 or more, but less than 10,000, car and/or rail passengers per day or 80 or more, but less than 2,000, ship passengers per day.
Megger [STD 8719.24 Annex]	high voltage resistance meter.
Meteoroids [STD 8719.14]	Naturally occurring particulates associated with solar system formation or evolution processes. Meteoroid material is often associated with asteroid breakup or material released from comets.
Milestone [STD 8729.1]	Any significant event in the program/project life cycle or in the associated reliability or maintainability program that is used as a control point for measurement of progress and effectiveness or for planning or redirecting future effort.
minor leak or spill [STD 8719.24 Annex]	a leak or spill that does not affect regions beyond the immediate work area, constitute a hazard to personnel, or involve damage to facilities or equipment; a minor leak or spill is less than one gallon.
Mishap [STD 8719.25]	Any unplanned event or series of events that results in death, injury, occupational illness, or damage to or loss of property.
mishap [STD 8719.24 Annex]	an unplanned event or series of events resulting in death, injury, occupational illness, or damage to or loss of equipment or property or damage to the environment.
Mishap Investigation Board [NPR 8621.1]	A NASA-sponsored board tasked to investigate the mishap or close call and to generate the mishap investigation report in accordance with the requirements specified in this NPR.

Term [Citing Document(s)]	Definitions
Mishap Investigation Report [NPR 8621.1]	The mishap investigation report documents the facts associated with an incident as determined by the investigating authority. In the report, the investigating authority identifies primary, or root, causes, and contributing and possible causes and recommends corrective actions to prevent the occurrence of similar mishaps.
Mishap Investigation Support Office, Mishap Regional Support Specialist [NPR 8621.1]	A NASA Safety Center Federal employee trained and experienced in all facets of NASA mishap investigation. Specialists assist and advise Centers, programs, projects, and investigating authorities on behalf of the Office of Safety and Mission Assurance Mishap Investigation Program Executive on implementation of policy and best-practice techniques to conduct and endorse NASA mishap and close call investigations.
Mishap Investigation Team [NPR 8621.1]	A NASA-sponsored team tasked to investigate a mishap or close call and generate the mishap investigation report in accordance with the requirements specified in this NPR.
Mishap Investigator [NPR 8621.1]	A Federal employee who serves as sole investigator for a mishap or close call and generates the mishap investigation report in accordance with the requirements specified in this NPR.
Mishap Preparedness and Contingency Plans [NPR 8621.1]	Pre-approved documents outlining timely organizational activities and responsibilities that must be accomplished in response to emergency, catastrophic, or potential (but not likely) events encompassing injuries, loss of life, property damage, or mission failure.
Mishap Summary [NPR 8621.1]	A formatted presentation prepared by the NASA Safety Center Mishap Investigation Support Office as a public-releasable document to capture the event sequence, findings, and recommendations contained in a NASA Type A, Type B, or high-visibility mishap or close call investigation report.
mismating [STD 8719.24 Annex]	the improper installation and/or connection of connectors.
Missile System Prelaunch Safety Package [STD 8719.24 Annex]	a data package demonstrating compliance with the system safety requirements of Volume 3, serves as a baseline for safety related information on the system throughout its life cycle; now known as the Safety Data Package (SDP).
Mission Abort [NPR 8705.2]	Same as "Abort." The forced early return of the crew to Earth when failures or hazards prevent continuation of the mission profile and a return to Earth is required to prevent a catastrophic event. The crew is safely returned to Earth in the space system nominally used for entry and landing/touchdown.
Mission Assurance [NPR 8715.3]	Providing increased confidence that applicable requirements, processes, and standards for the mission are being fulfilled.
Mission Assurance Process Map [NPR 8705.6]	The mission assurance process map is a high-level, graphical representation of governing SMA policy and requirements, processes, and key participant roles, responsibilities, and interactions. It also includes the reporting structure that constitutes a program's/project's SMA functional flow.

Term [Citing Document(s)]	Definitions
Mission Assurance Process Matrix [NPR 8705.6]	The mission assurance process matrix is constructed to identify program life cycle assurance agents and specific assurance activities, processes, responsibilities, accountability, depth of penetration, and independence. The matrix includes key assurance personnel in Engineering, Manufacturing, Program Management, Operations, and SMA.
Mission Critical [NPR 8715.3]	Item or function that must retain its operational capability to assure no mission failure (i.e., for mission success).
Mission Critical [STD 8729.1]	[1] Item or function that must retain its operational capability to assure no mission failure (i.e., for mission success). [2] An item or function, the failure of which may result in the inability to retain operational capability for mission continuation if a corrective action is not successfully performed.
Mission Essential Personnel [NPR 8715.5]	Government or contractor personnel who are directly involved in ensuring the safety and success of a mission. For the purposes of range flight safety, mission essential personnel do not include any people on board the vehicle.
Mission Essential Personnel [STD 8719.25]	Government or contractor personnel who are directly involved in ensuring the safety and success of a mission. For the purposes of range safety, mission essential personnel do not include any personnel on board the vehicle.
Mission Failure [NPR 8621.1]	A mishap of whatever intrinsic severity prevents the achievement of the mission's minimum success criteria or minimum mission objectives as described in the mission operations report or equivalent document.
Mission Hardware [STD 8739.6]	Hardware used in Category 1 and Category 2 projects and/or Class A, B, or C payloads, including critical support hardware.
Mission Operations [STD 8719.14]	All activities executed by the spacecraft; includes design mission, primary mission, secondary mission, extended mission, and disposal.
Mission Success [NPR 8715.3]	Meeting all mission objectives and requirements for performance and safety.
Mitigation [STD 8729.1]	An action taken or planned to reduce the consequence of an event (synonyms: compensating provisions, fault-tolerance).
Mix Record [STD 8739.1]	A record of the procedure followed for mixing the polymeric compounds.
Mobile Aerial Platform [STD 8719.9]	A mobile device that has an adjustable position platform and is supported from ground level by a structure.
Mode [STD 8739.5]	In general, an electromagnetic field distribution that depends on wavelength of light and material properties of the traveling medium. In guided wave propagation, such as through a waveguide or optical fiber, a distribution of electromagnetic energy that satisfies Maxwell's equations and boundary conditions. In terms of ray optics, a possible path followed by light rays dependent on index of refraction, wavelength of light and waveguide dimensions.

Term [Citing Document(s)]	Definitions
Moderator [STD 8739.9]	The individual who leads an inspection. Responsible for planning the inspection events with the author, scheduling and facilitating meetings, collecting and reporting measurements from inspections he/she moderates, and possibly verifying the author's rework. (see Wiegers 2008)
Module [STD 8739.1]	A separable unit in a packaging scheme.
Module [STD 8739.9]	A program unit that is discrete and identifiable with respect to compiling, combining with other units, and loading; for example, input to, or output from, an assembler, compiler, linkage editor, or an executive routine. (see IEEE Std 610.12-1990)
Molding [STD 8739.4]	The sealing of a connector backshell area or a cable breakout with a compound or material that excludes moisture and provides stress relief. The material is injected into molds that control its configuration.
monitor circuit [STD 8719.24 Annex]	a circuit used to verify the status of a system, such as an inhibit directly; control circuits can be monitored but they cannot serve as a monitor circuit.
Multi-mode Fiber [STD 8739.5]	An optical fiber that will allow two or more bound modes to propagate in the core at the wavelengths of interest.
Nadcap [NPR 8735.2]	An aerospace industry third party accreditation program which conducts supplier audits and provides accreditation/certification that a supplier is competent to furnish a specified product, process, or service. Nadcap program requirements, including the criteria, terms, and governance structure for Nadcap accreditation, are provided in SAE AS7003. Details regarding Nadcap accreditation services, a list of processes for which NADCAP provides supplier accreditation, and Nadcap auditing attributes can be found at http://www.prinetwork.org/NADCAP/.
NASA Advisories, Notices, and Alerts Distribution and Response Tracking System (NANADARTS) [NPR 8735.1]	A Web-based tool operated by NASA and used to distribute NASA Advisories, GIDEP Notices, and other related documents and collect closed-loop responses for these documents.
NASA Advisory [NPR 8735.1]	A NASA document for exchanging significant parts, materials, and safety problems or concerns among NASA activities.
NASA Advisory Reporting System (NARS) [NPR 8735.1]	A web-based tool operated by NASA and used to generate, document and store NASA Advisories.
NASA Aircraft [NPR 8621.1]	Aircraft that are bought, borrowed, chartered, rented, or otherwise procured or acquired— including aircraft produced with the aid of NASA funding—regardless of cost, from any source for the purpose of conducting NASA science, research, or other missions, and which are NASA-operated or NASA-managed. Unmanned aircraft are defined as "aircraft" by the Federal Aviation Administration and are included in the definition of NASA aircraft unless specified otherwise.

Term [Citing Document(s)]	Definitions
NASA Contractor or Grantee Mishap or Close Call [NPR 8621.1]	A mishap or close call requiring a NASA contractor or grantee to report or investigate it according to provisions in the contactor or grantee's contract.
NASA Controlled Range Flight Operations [NPR 8715.5]	These are operations from: 1) a NASA range, or an offsite range where NASA is the range authority for the operation; (e.g., KSC; WFF; Kodiak, Alaska) 2) operations by a NASA-operated or controlled vehicle; or 3) operations involving a NASA crew or payload which are not FAA-licensed.
NASA Controlled Range Flight Operations [STD 8719.25]	These are operations: 1) from a NASA range, or an offsite range where NASA is the range authority for the operation (e.g. KSC, WFF, or Kodiak, AK); 2) by a NASA operated or controlled vehicle; or 3) involving a NASA crew or payload which are not FAA licensed.
NASA Employees [NPR 8621.1]	Federal civil servants employed and paid by NASA, or on detail from other Federal agencies, and NASA Support Service Contractors.
NASA Human Spaceflight Missions [NPR 8705.2]	Terminology used to distinguish human spaceflight missions that require human-rated systems per this NPR. Any human spaceflight mission where NASA retains the mission decision authority and the responsibility for crew safety is considered a NASA mission.
NASA Level A Training Center [STD 8739.6]	The Eastern NASA Manufacturing Technology Transfer Center (E-NMTTC) at NASA Goddard Space Flight Center and the Western NASA Manufacturing Technology Transfer Center (W-NMTTC) at the Jet Propulsion Laboratory are NASA Level A training centers.
NASA Mishap [NPR 8621.1]	A NASA mishap is an unplanned event resulting in at least one of the following: a. Occupational injury or occupational illness to non-NASA personnel caused by NASA operations. b. Occupational injury or occupational illness to NASA personnel caused by NASA operations. c. Destruction of or damage to NASA property, public or private property, including foreign property, caused by NASA operations or NASA-funded research and development projects. d. NASA mission failure before the scheduled completion of the planned primary mission.
NASA Mishap Information System [NPR 8621.1]	A custom-developed system for capturing mishaps, close calls, and hazards, as required in this NPR.
NASA Operation [NPR 8621.1]	An activity or process under direct NASA physical, administrative, or contractual control or where significant NASA resources are dedicated to accomplishing an objective common to NASA and other independent organizations. This does not include non-NASA contracted or funded activities conducted at a common location or environment with NASA resources.
NASA Operation [STD 8719.9]	Any activity or process under NASA direct control or that includes major NASA involvement.

Term [Citing Document(s)]	Definitions
NASA Safety Standard (NSS) [NPR 8715.3]	A NASA safety document that requires conditions, or the adoption or use of one or more practices, means, methods, operations, or processes reasonably necessary or appropriate to provide for safe employment and places of operation. The document is promulgated by the NASA Office of Safety and Mission Assurance and implemented and enforced by the Center Safety and Mission Assurance organizations.
NASA Workforce [NPR 8715.5]	Government and contractor personnel who are directly involved in a range flight operation or who work at a range, launch site, or landing site where a NASA range flight operation takes place. For the purposes of this NPR, "workforce" does not include any crew on board a vehicle during flight.
NASA Workforce [STD 8719.25]	Government and contractor personnel who are directly involved in a range flight operation or who work at a range, launch site, or landing site where a NASA range flight operation takes place. For the purposes of this standard, "workforce" does not include any crew on board a vehicle during flight.
NASA Workmanship Standards Technical Committee [STD 8739.6]	NASA civil service employees who are the primary points of contact for the NASA Workmanship Standards Program for each NASA Center. See http://nepp.nasa.gov/workmanship for the current roster.
National Airspace System (NAS) [NPR 8715.5] [STD 8719.25]	The common network of U.S. airspace controlled by the FAA including air navigation facilities, equipment and services, airports or landing areas, aeronautical charts, information and services, rules, regulations, and procedures, technical information, and manpower and material. Also included are system components shared jointly with the military.
National Transportation Safety Board Serious Injury [NPR 8621.1]	Any injury resulting from an aircraft mishap in which one or more of the following apply: a. Requires hospitalization for more than 48 hours, commencing within seven days from the date the injury was received. b. Results in a fracture of any bone except for simple fractures of fingers, toes, or nose. c. Causes severe hemorrhages or nerve, muscle, or tendon damage. d. Involves any internal organ. e. Involves second- or third-degree burns or any burns affecting more than five percent of the body surface.
nationally recognized testing laboratory [STD 8719.24 Annex]	see testing laboratory (nationally recognized).
Net Explosive Weight (NEW) [STD 8719.12]	The total quantity, expressed in pounds, of explosive material or pyrotechnics in a container or device. The NEW may include the mass of the TNT-equivalent of all contained energetic substances based on Center policies.
Net Explosive Weight for Quantity Distance (NEWQD) [STD 8719.12]	The total quantity, expressed in pounds, of high explosive (HE) equivalency in each item to be used when applying quantity-distance criteria. The NEWQD is equal to the NEW unless hazard classification testing has shown that a lower weight is appropriate for Quantity Distance (QD) purposes.

Term [Citing Document(s)]	Definitions
no-fire level [STD 8719.24 Annex]	the maximum direct current or radio frequency energy at which an electroexplosive initiator shall not fire with a reliability of 0.999 at a confidence level of 95 percent as determined by a Bruceton test and shall be capable of subsequent firing within the requirements of performance specifications.
Non-applicable Requirement [STD 8709.20]	Not relevant, not capable of being applied (from NASA Memo 7120- 81, Appendix A, which updated NPR 7120.5D).
Non-Code PVS [STD 8719.17]	Any pressure vessel that is not stamped with the appropriate symbol and documented as complying with the original applicable construction Code or any pressure piping system that does not meet the requirements of the appropriate fabrication code (e.g. ASME Section VIII, B31.1, B31.3), including PVS that were fabricated from non-Code materials by non-Code processes or organizations.
Noncombustible [STD 8719.11]	Structures in which the structure itself (exclusive of trim, interior finish, and contents) is noncombustible but not fire-resistive. Common forms include exposed steel beams and columns, and masonry or metal walls.
Non-common conductors [STD 8739.1]	Conductive surfaces that are not attached to the same electric node.
Noncompliance [NPR 8715.7]	An instance of failure to satisfy a requirement.
noncompliance [STD 8719.24 Annex]	a noticeable or marked departure from requirements, standards, or procedures; includes equivalent level of safety determinations (formerly meets intent certifications), and waivers.
Nonconformance [NPR 8735.1]	The state or situation of not fulfilling a requirement. A nonconforming product, process, software, or material does not meet manufacturing specifications or design, composition, or contractual requirements. Counterfeit parts, products, software, and materials are considered nonconforming and/or nonconformances.
noncritical hardware [STD 8719.24 Annex]	equipment and systems used for standard industry use; equipment or systems that are determined not to be hazardous, of high value, or safety critical.
Noncritical Lift [STD 8719.9]	A lift involving routine lifting operations governed by standard industry rules and practices except as supplemented with unique NASA testing, operations, maintenance, inspection, and personnel licensing requirements contained in this standard.
Nondestructive Evaluation (NDE) [STD 8719.9]	See Nondestructive Testing.
Nondestructive Examination [STD 8719.17]	The application of technical methods to examine materials or components in ways that do not impair future usefulness and serviceability in order to detect, locate, measure, and evaluate flaws; to assess integrity, properties, and composition; and to measure geometrical characteristics.

Term [Citing Document(s)]	Definitions
Nondestructive Testing (NDT) [STD 8719.9]	The application of technical methods to examine materials or components in ways that do not impair future usefulness and serviceability in order to detect, locate, measure, and evaluate flaws; to assess integrity, properties, and composition; and to measure geometrical characteristics.
nonessential personnel [STD 8719.24 Annex]	those persons not deemed launch-essential or neighboring operations personnel; includes the general public, visitors, the media, and any persons who can be excluded from Safety Clearance Zones with no effect on the operation or parallel operations.
Non-essential Personnel [STD 8719.12]	Personnel not essential to, or involved with, the immediate operation presenting the energetic materials hazard.
non-incendive [STD 8719.24 Annex]	will not ignite group of gases or vapors for which it is rated. Similar to intrinsically safe, but does not include failure tolerance ratings; used in rating electrical products for Class I, Division 2 locations only.
Non-Load Test Slings, Rigging Hardware, And Below-The-Hook Lifting Devices [STD 8719.9]	Slings, rigging hardware, and below-the hook lifting devices meeting the criteria of section 14.5.4 and designated and approved by the LDEM as not subject to periodic load testing requirements.
Non-mass Explosion [STD 8719.12]	Partial explosion of a mass of explosives when only a small portion is subjected to fire, severe concussion or impact, the impulse of an initiating agent, or to the effect of a considerable discharge of energy from an outside stimulus. Also refers to sequential propagation of explosions of multiple items with time delays such that blast overpressure effects do not combine from each individual explosion.
Nuclear Flight Safety Assurance Manager (NFSAM) [NPR 8715.3]	The person in the Office of Safety and Mission Assurance responsible for assisting the project offices in meeting the required nuclear launch safety analysis/evaluation.
Objective Evidence [NPR 8705.6]	Data verifying or supporting the existence of a finding or compliance to requirements. Objective evidence may be obtained through observation, measurement, test, or other means and is not influenced by prejudice, emotion, or bias. Examples of objective evidence include, but are not limited to, procedures, records, work instructions, databases, reports, organizational charts, interviews, hardware, facilities test reports, configuration control documentation (i.e., drawings and specifications), mishap reports, corrective actions, and lessons learned.
Observation [NPR 8621.1]	A factor, event, or circumstance identified during an investigation that did not contribute to the mishap or close call, but if left uncorrected, has the potential to cause a mishap or increase the severity of a mishap; or a positive factor, event, or circumstance that should be noted.
Obsolete Part [STD 8739.10]	A part that is no longer being manufactured.
Occupational Injury or Illness	Work-related per 29 CFR pt. 1904.

Term [Citing Document(s)]	Definitions
[NPR 8621.1]	
Occupational Safety and Health Administration (OSHA) [NPR 8715.3]	The Federal agency which promulgates and enforces workplace safety regulations and guidance.
Occupational Safety and Health Administration Final Mishap Summary [NPR 8621.1]	A report (OSHA 301 Form: Injury and Illness Incident Report, or an equivalent form) provided in accordance with 29 CFR pt. 1960.70 by NASA to the Office of Federal Agency Programs for each mishap involving an OSHA-recordable incident.
Occupied Facility [STD 8719.11]	A building or facility occupied by persons on a regular basis and not used for sleeping purposes.
Off The Shelf [STD 8739.6]	Products sold in the common marketplace, without modification, that are made and procured to a supplier-defined design, supplier-defined form, fit and function specifications, and supplier-defined quality assurance requirements.
Office of the Chief of Safety [STD 8719.24 Annex]	the range office headed by the Chief of Safety; this office ensures that the Range Safety Program meets range and Range User needs and does not impose undue or overly restrictive requirements on a program.
Off-The-Shelf (OTS) Software [STD 8739.9]	Software not developed for the specific project now underway. The software is considered general purpose or developed for a different project or projects. If Commercially created - COTS; created by another government source - GOTS; Modified prior to use - MOTS. OTS may include legacy, heritage, and re-use software.
Off-The-Shelf (OTS) software [STD 8719.13]	Includes Commercial Off-The-Shelf (COTS), Government Off-The-Shelf (GOTS), and Modified Off-The-Shelf (MOTS) software. OTS software may also be legacy, heritage, and re-use software. Refer to section 6.6 of this Standard for applicability to COTS.
Off-The-Shelf Hardware [STD 8739.10]	Assembly, part, or design that is readily available for procurement, usually to catalog specifications, without the necessity of generating detail procurement specifications for the item.
Open Plan [STD 8719.11]	When referring to office space, it denotes large floor areas (greater than 3,000 square feet [279 square meters]) characterized by the lack of fixed, ceiling-high partitions and conventional doorways. Individual workstations are identified by the arrangement of desks, chairs, files, bookcases, and movable partitions. The hazard from a fire safety standpoint is due to the ill-defined nature of means of egress and the lack of a significant physical barrier against the spread of smoke and fire, thus magnifying potential loss.
Operability [NPR 8715.3]	As applied to a system, subsystem, component, or device is the capability of performing its specified function(s) including the capability of performing its related support function(s).
Operating Building [STD 8719.12]	Any structure, except a magazine, in which operations pertaining to manufacturing, processing, or handling explosives are performed.

Term [Citing Document(s)]	Definitions
operating environment [STD 8719.24 Annex]	an environment that a launch vehicle component will experience during acceptance testing, launch countdown, and flight; includes shock, vibration, thermal cycle, acceleration, humidity, and thermal vacuum.
Operating Line [STD 8719.12]	Group of buildings used to perform the consecutive steps in the loading, assembling, modification, or salvaging of an item or in the manufacture of an explosive or explosive device.
operation [STD 8719.24 Annex]	a scheduled activity where range assets are necessary to support Range User requirements for a specified time period.
Operational Readiness [STD 8729.1]	The ability of a system to respond and perform its mission upon demand.
Operational Safety [NPR 8715.3]	That portion of the total NASA safety program dealing with safety of personnel and equipment during launch vehicle ground processing, normal industrial and laboratory operations, use of facilities, special high hazard tests and operations, aviation operations, use and handling of hazardous materials and chemicals from a safety viewpoint.
Operational Safety [NPR 8715.7]	That portion of the total NASA safety program dealing with safety of personnel and equipment during launch vehicle ground processing, normal industrial and laboratory operations, use of facilities, special high hazard tests and operations, aviation operations, and use and handling of hazardous materials and chemicals from a safety viewpoint.
Operational Shield [STD 8719.12]	A barrier constructed to protect personnel, material, or equipment from the effects of a possible fire or explosion occurring at a particular operation.
operations safety plan [STD 8719.24 Annex]	the detailed safety procedures used for missile operations; these plans are written by the Range Contractor and Operations Safety; includes Explosives Safety Plans, Facility Safety Plans, and Safety Operational Plans.
Operator [NPR 8705.2]	Any human interacting with the crewed space system during the mission.
Operator [STD 8719.12]	A person assigned to perform a specific, generally continuing function on a production, maintenance, or disposal line or operation. Typically, the functions are performed at workstations or areas defined in a Standard Operating Procedure (SOP).
optical coverage ratio [STD 8719.24 Annex]	the percentage of the surface area of the cable core insulation covered by a shield.
Optical Time Domain Reflectometry (OTDR) Backscattering Technique [STD 8739.5]	A method for characterizing an optical fiber whereby an optical pulse is transmitted through the fiber and the optical power of the resulting light scattered and reflected back to the input is measured as a function of time.
Orbital Debris [STD 8719.14]	Artificial objects, including derelict spacecraft and spent launch vehicle orbital stages, left in orbit which no longer serve a useful purpose. In this document, only debris of diameter 1 mm and larger is considered. If liquids are to be released, they should explicitly be shown to be compliant with all mitigation requirements.

Term [Citing Document(s)]	Definitions
Orbital Insertion [NPR 8715.5]	With regard to the application of requirements and criteria in this NPR to a space launch, orbital insertion occurs when the vehicle or component achieves a minimum 70 nm perigee based on a computation that accounts for drag.
Orbital Insertion [STD 8719.25]	With regard to the application of requirements and criteria in this standard to a space launch, orbital insertion occurs when the vehicle or component achieves a minimum 70 nm perigee based on a computation that accounts for drag.
Orbital Lifetime [STD 8719.14]	The length of time an object remains in orbit. Objects in LEO or passing through LEO lose energy as they pass through the Earth's upper atmosphere, eventually getting low enough in altitude that the atmosphere removes them from orbit.
Orbital Stage [STD 8719.14]	A part of the launch vehicle left in a parking, transfer, or final orbit during or after payload insertion; includes liquid propellant systems, solid rocket motors, and any propulsive unit jettisoned from a spacecraft.
Ordinary [STD 8719.11]	Masonry exterior load-bearing walls or load-bearing portions of exterior walls that are of noncombustible construction.
Ordnance [STD 8719.12]	Explosives, chemicals, pyrotechnics, and similar stores (e.g., bombs, guns and ammunition, flares, smoke, or napalm). The term is sometimes used interchangeably with "explosives".
ordnance [STD 8719.24 Annex]	all ammunition, demolition material, solid rocket motors, liquid propellants, pyrotechnics, and explosives as defined in AFMAN 91-201 and DoD 6055.9-STD.
ordnance component [STD 8719.24 Annex]	a component such as a squib, LOS, detonator, initiator, igniter, or linear shape charge in an ordnance system.
ordnance operation [STD 8719.24 Annex]	any operation consisting of shipping, receiving, transportation, handling, test, checkout, installation and mating, electrical connection, render safe, removal and demating, disposal, and launch of ordnance.
Organizational Factor [NPR 8621.1]	Any operational or management structural entity that exerts control over the system at any stage in its life cycle including, but not limited to, the system's concept development, design, fabrication, test, maintenance, operation, and disposal—for example, resource management (budget, staff, training); policy (content, implementation, verification); and management decisions.
Organizational Unit [NPR 8000.4]	An organization, such as a program, project, Center, Mission Directorate, or Mission Support Office that is responsible for carrying out a particular activity.
OSMA Policy/Requirements Website [STD 8709.20]	Website where OSMA documents can be accessed: http://www.hq.nasa.gov/office/codeq/doctree/index.htm
Outgassing [STD 8739.4]	The release of a volatile part(s) from a substance when placed in a vacuum environment.
Override	To take precedence over system control functions.

Term [Citing Document(s)]	Definitions
[NPR 8705.2]	
Oversight [STD 8739.8]	Surveillance mode that is in line with the supplier's processes. The acquirer retains and exercises the right to concur or non-concur with the supplier's decisions. Non-concurrence must be resolved before the supplier can proceed. Oversight is a continuum that can range from low intensity, such as acquirer concurrence in reviews (e.g., PDR, CDR), to high intensity oversight, in which the customer has day-to-day involvement in the supplier's decision-making process (e.g., software inspections).
Oversight/Insight [NPR 8715.3]	The transition in NASA from a strict compliance-oriented style of management to one which empowers line managers, supervisors, and employees to develop better solutions and processes.
Owner [STD 8719.17]	The management of the organization responsible for the PVS as defined in NPD 8710.5, Policy for Pressure Vessels and Pressurized Systems.
Oxidizer [STD 8719.12]	A chemical (other than a blasting agent or explosive as defined in 29 CFR 1910.109(a) that initiates or promotes combustion in other materials, thereby causing fire either of itself or through the release of oxygen or other gases.
Part Lead [STD 8739.1]	The conductor attached to an electrical, electronic or electromechanical (EEE) part.
Partial Traceability [STD 6008]	Documentation from a supplier or vendor that does not necessarily include the full chain of custody back to the original fastener manufacturer.
Passenger [NPR 8705.2]	Any human on board the space system while in flight that has no responsibility to perform any mission task for that system. Often referred to as "Space Flight Participant."
Passivation [STD 8719.14]	The process of removing stored energy from a space structure at EOM which could result in an explosion or deflagration of the space structure to preclude generation of new orbital debris after End of Mission. This includes removing energy in the form of electrical, pressure, mechanical, or chemical.
passive component [STD 8719.24 Annex]	a flight termination system component that does not contain active electronic piece parts such as microcircuits, transistors, and diodes: includes, but need not be limited to, radio frequency antennas, radio frequency couplers, and cables and rechargeable batteries, such as nickel cadmium batteries.
Payload [NPR 8715.5]	The object(s) within a payload fairing carried or delivered by a vehicle to a desired location or orbit.
Payload [NPR 8715.7]	The object(s) within a payload fairing carried or delivered by a launch vehicle to a desired location or orbit includes but is not limited to satellites, other spacecraft, experimental packages, reentry vehicles, dummy loads, cargo, and any motors attached to them in the payload fairing.
Payload [STD 8719.25]	The object(s) carried or delivered by a vehicle to a desired location or orbit.

Term [Citing Document(s)]	Definitions
Payload [NPR 8735.1]	Any airborne or space equipment or material that is not an integral part of the carrier vehicle (i.e., not part of the carrier aircraft, balloon, sounding rocket, expendable or recoverable launch vehicle). Included are items such as free-flying automated spacecraft, Space Shuttle payloads, Space Station payloads, Expendable Launch Vehicle payloads, flight hardware and instruments designed to conduct experiments, and payload support equipment. [NPR 8705.4 Risk Classification for NASA Payloads]
payload [STD 8719.24 Annex]	the object(s) within a payload fairing carried or delivered by a launch vehicle to a desired location or orbit.
payload processing facility and launch site area [STD 8719.24 Annex]	the areas and support facilities (such as payload processing facilities and launch pad) where the payload is processed, stored, or transported in support of final payload processing, payload to launch vehicle integration, and launch.
Payload Safety Introduction Briefing [NPR 8715.7]	The first meeting of a payload project's PSWG where the Payload Project briefs the payload to the safety community. This meeting is also referred to as the Concept Briefing with respect to AFSPCMAN 91-710, Range Safety User Requirements.
Payload Safety Working Group [NPR 8715.7]	A working group formed for each NASA ELV payload with a primary purpose to ensure (1) a project's compliance with applicable safety requirements and (2) that the safety risk is identified, understood, and adequately controlled (see paragraph 2.2 of this NPR).
Payload Safety Working Group [STD 8719.24 Annex]	A working group formed for each NASA ELV payload with a primary purpose to (1) ensure a project's compliance with applicable safety requirements and (2) that the safety risk is identified, understood, and adequately controlled.
Peeling [STD 8739.1]	The separation of conformal coating from the PWA, usually due to improper preparation or abrasion. Peeling is distinguished from lifting in that the layer of conformal coating is not continuous.
Peer Review [STD 8739.8]	A review of a software work product, following defined procedures, by peers of the producers of the product for the purpose of identifying defects and improvements. [SEI-CMM Software Engineering Institute Capability Maturity Model®]
Peer Review [STD 8739.9]	[1] A review of a software work product, following defined procedures, by peers of the producers of the product for the purpose of identifying defects and improvements. [2] Independent evaluation by internal or external subject matter experts who do not have a vested interest in the work product under review. Peer reviews can be planned, focused reviews conducted on selected work products by the producer's peers to identify defects and issues prior to that work product moving into a milestone review or approval cycle.
Penetration Debris Flux [STD 8719.14]	The number of impacts per square meter per year that will penetrate a surface of specified orientation with specified materials and structural characteristics.
Performance [STD 8739.9]	A measure of how well a system or item functions in the expected environments.
Performance Measure [NPR 8000.4]	A metric used to measure the extent to which a system, process, or activity fulfills its intended objectives.

Term [Citing Document(s)]	Definitions
Performance Requirement [NPR 8000.4]	The value of a performance measure to be achieved by an organizational unit's work that has been agreed upon to satisfy the needs of the next higher organizational level.
performance specification [STD 8719.24 Annex]	a statement prescribing the particulars of how a component or part is expected to perform in relation to the system that contains the component or part; includes specific values for range of operation, input, output, or other parameters that define the component's or part's expected performance.
Perigee [STD 8719.14]	The point in the orbit that is nearest to the center of the Earth. The perigee altitude is the distance of the perigee point above the surface of the Earth.
Periodic Inspection [STD 8719.9]	A thorough examination of LDE conducted at predetermined intervals (typically monthly to yearly) to assess the condition of the equipment. These inspections do not include pre-use inspections performed each day before the equipment is used. Details of these inspections are provided in regulations and industry standards.
Periodic Load Test [STD 8719.9]	A load test performed at predetermined intervals to determine whether the equipment (e.g., limit switches, E-Stop, controls, brakes, slings, shackles) is functioning properly.
Permanent Disability [NPR 8705.2]	A non-fatal occupational injury or illness resulting in permanent impairment through loss of, or compromised use of, a critical part of the body, to include major limbs (e.g., arm, leg), critical sensory organs (e.g., eye), critical life-supporting organs (e.g., heart, lungs, brain), and/or body parts controlling major motor functions (e.g., spine, neck). Therefore, permanent disability includes a non-fatal injury or occupational illness that permanently incapacitates a person to the extent that he or she cannot be rehabilitated to achieve gainful employment in their trained occupation and results in a medical discharge from duties or civilian equivalent.
Permanent Partial Disability [NPR 8621.1]	An injury or occupational illness that does not result in a fatality or permanent total disability, but in the opinion of competent medical authority, results in permanent impairment through loss of use of any body part with the following exceptions: loss of teeth, fingernails, or toenails; loss of tip of fingers or toes without bone involvement; inguinal hernia (if it is repaired); disfigurements; or sprains or strains that do not cause permanent limitation of motion.
Permanent Total Disability [NPR 8621.1]	A nonfatal injury or occupational illness that, in the opinion of competent medical authority, permanently and totally incapacitates a person to the degree where he or she cannot follow any gainful occupation and results in a medical discharge or civilian equivalent.
Personnel Access Platform [STD 8719.9]	A platform, typically deployed or relocated by one or multiple dedicated hoists or winches, which allow personnel to access and work in a specific area of a fixed structure or building. Personnel occupy these platforms only after the platforms are deployed and secured and never during movement or while the platforms are supported by hoists/winches. For platforms specifically designed to lift and lower persons via a hoist/winch, refer to Hoist-Supported Personnel Lifting Devices.
Personnel Access Platform Hoist/Winch	A dedicated hoist/winch whose only purpose is to raise and lower a personnel access platform not carrying personnel.

Term [Citing Document(s)]	Definitions
[STD 8719.9]	
Personnel Licensing [STD 8719.9]	A means to ensure an individual is qualified to perform a designated task.
personnel work platforms [STD 8719.24 Annex]	platforms used to provide personnel access to flight hardware at off-pad processing facilities as well as at the launch pad; they may be removable, extendible, or hinged.
Phase [STD 8739.9]	The period of time during the life cycle of a project in which a related set of software engineering activities is performed. Phases may overlap.
Physical Configuration Audit (PCA) [STD 8739.8]	An audit conducted to verify that one or more configuration items, as built, conform to the technical documentation that defines it. [Based on IEEE 610.12, IEEE Standard Glossary of Software Engineering Terminology]
Pistoning [STD 8739.5]	The axial movement of an optical fiber within a connector or connector ferrule.
pneumatic [STD 8719.24 Annex]	operated by air or other gases under pressure.
Policy Waiver [STD 8719.17]	Documented and approved Center policy contrary to the policy in NPD 8710.5, Policy for Pressure Vessels and Pressurized Systems, or this standard. (See NPR 8715.3, paragraph 1.13, for waiver requirements.)
Polymer [STD 8739.1]	A compound of high molecular weight that is derived from either the joining together of many small similar or dissimilar organic molecules or by the condensation of many small molecules by the elimination of water, alcohol, or a solvent.
populated area [STD 8719.24 Annex]	an outdoor location, structure, or cluster of structures that may be occupied by people; sections of roadways and waterways that are frequented by automobile and boat traffic are populated areas; agricultural lands, if routinely occupied by field workers, are also populated areas.
positive control [STD 8719.24 Annex]	the continuous capability to ensure acceptable risk to the public is not exceeded throughout each phase of powered flight or until orbital insertion.
Postmission Disposal [STD 8719.14]	The orbit/location where a spacecraft/launch vehicle is left after passivation at EOM.
Pot Life (aka. Working Life) [STD 8739.1]	The length of time a material, substance, or product is in use while it meets all applicable requirements and remains suitable for its intended use (IPC T-50).
Potential Explosive Site (PES) [STD 8719.12]	Location of a quantity of explosives that will create a blast fragment, thermal, or debris hazard in the event of an accidental explosion of its contents. The distance to an ES determines quantity limits for ammunition and explosives at a PES.
power source [STD 8719.24 Annex]	(1) a battery; (2) the point of direct current (DC) to alternating current (AC) conversion for capacitor charged systems.
Precursor	An occurrence of one or more events that have significant failure or risk implications.

Term [Citing Document(s)]	Definitions
[NPR 8715.3]	
Preliminary Hazard Analysis [STD 8719.13]	A gross study of the initial system concepts. It is used to identify all of the energy sources that constitute inherent hazards. The energy sources are examined for possible accidents in every mode of system operation. The analysis is also used to identify methods of protection against all of the accident possibilities. Software's high level roles in contributing to or protecting the system should be considered and recorded (e.g. software's inadvertent release of an energy source or the detection and inhibit of an energy source).
pressure component [STD 8719.24 Annex]	a component such as lines, fittings, valves, regulators, and transducers in a pressurized system; normally pressure vessels or pressurized structures are excluded, because of the potential energy contained; they generally require additional analysis, test and inspection.
Pressure Relief Device (PRD) [STD 8719.17]	A device designed to open without intervention by an operator and relieve excess pressure so as to protect the PVS on which it is installed from damage due to that pressure.
Pressure Relief Valve (PRV) [STD 8719.17]	A pressure relief device designed to actuate on inlet static pressure and to reclose after normal conditions have been restored.
pressure system [STD 8719.24 Annex]	any system above 0 psig that is classified as follows: low pressure, 0 to 500 psi; medium pressure, 501 to 3000 psi; high pressure, 3001 to 10,000 psi; ultra-high pressure, above 10,000 psi. The degree of hazard of a pressure system is proportional to the amount of energy stored, not the amount of pressure it contains; therefore, low pressure, high volume systems can be as hazardous to personnel as high pressure systems; see pressurized system.
Pressure Systems Manager [STD 8719.17]	The person responsible for implementation of NPD 8710.5, Policy for Pressure Vessels and Pressurized Systems, and this standard at a NASA facility.
Pressure Vessel [NPR 8715.3]	Any vessel used for the storage or handling of a fluid under positive pressure. A pressure system is an assembly of components under pressure; e.g., vessels, piping, valves, relief devices, pumps, expansion joints, gages.
pressure vessel [STD 8719.24 Annex]	a container that stores pressurized fluids and (1) contains stored energy of 14,240 foot pounds (19,130 joules) or greater based on adiabatic expansion of a perfect gas; or (2) contains gas or liquid which will create a mishap (accident) if released; or (3) will experience a MEOP greater than 100 psia; excluded are special equipment including batteries, cryostats (or dewars), heat pipes, and sealed containers; or (4) per the ASME definition, summarized briefly; pressure containers that are integral pumps or compressors, hot water heaters and boilers, vessels pressurized in excess of 15 psi (regardless of size), and vessels with a cross-sectional dimension greater than 6 inches (regardless of length of the vessel or pressure).
Pressure Vessels and Pressurized Systems (PVS) [STD 8719.17]	Pressure vessels and pressurized systems within the scope of NPD 8710.5, Policy for Pressure Vessels and Pressurized Systems, and this standard.

Term [Citing Document(s)]	Definitions
pressurized structure [STD 8719.24 Annex]	a structure designed to carry both internal pressure and vehicle structural loads; the main propellant tank of a launch vehicle is a typical example.
pressurized system [STD 8719.24 Annex]	a system that consists of pressure vessels or pressurized structures, or both, and other pressure components such as lines, fittings, valves, and bellows that are exposed to and structurally designed largely by the acting pressure; electrical or other control devices required for system operation are not included; a pressurized system is often called a pressure system; see pressure system.
Prevention [STD 8729.1]	An action taken to reduce the likelihood of an event (Synonyms: preventive measure, fault avoidance).
Prime Contractor [STD 6008]	A contractor who has been given responsibility through NASA to manage a major flight-level program that may involve development through design, manufacture, testing and integration, launch, and post-launch activities.
Printed Circuit Board (PCB) [STD 8739.1]	A composite structure incorporating point-to-point interconnections for electronic circuits. It may include embedded components. (This includes single-sided, double-sided, multi-layer, rigid, rigid-flex, and flex constructions (IPC T-50)).
Printed Wiring Assembly (PWA) [STD 8739.1]	The PWA consists of the PCB, components, and associated hardware and materials.
Probabilistic Risk Assessment (PRA) [NPR 8715.3]	A PRA is a comprehensive, structured, and logical analysis method aimed at identifying and assessing risks in complex technological systems for the purpose of cost-effectively improving their safety and performance in the face of uncertainties. PRA assesses risk metrics and associated uncertainties relating to likelihood and severity of events adverse to safety or mission.
Probabilistic Safety Requirement [NPR 8705.2]	The specification of a criterion for a probabilistic safety metric (e.g., the probability of a loss of crew) and the degree of certainty with which such criteria must be met.
Probability of Casualty (Pc) [STD 8719.25]	A measure of individual risk. Pc is the probability that an individual at a specific location would be a casualty per an event, such as vehicle flight, if a large number of events could be carried out under identical circumstances. For example, if an individual would be a casualty once per one million identical launches, the Pc for such a launch would be 1×10-6.
Probability of Impact (Pi) [STD 8719.25]	The probability that one or more pieces of debris from a vehicle will impact a given location or object (e.g., aircraft, ships).
Problem/Failure/Anomailies Management (P/F/A) [STD 8729.1]	A formalized process to document, resolve, verify, correct, review and archive P/F/A incurred during the development of functional hardware or software.
Procedure [NPR 8621.1]	A documented description of the sequential actions in performing a given task.
Process [NPR 8621.1]	A set of activities used to convert inputs into desired outputs to generate expected outcomes and satisfy a purpose.

Term [Citing Document(s)]	Definitions
Process [STD 8739.8]	A set of interrelated activities, which transform inputs into outputs. [ISO/IEC 12207, Software life cycle processes]
Process Assurance [STD 8739.8]	Activities to assure that all processes involved with the project adhere to plans and comply with the contract and/or any memorandum of agreement/understanding.
Process Witnessing [NPR 8735.2]	Physical observation of a contractor test or work process to ensure that the process is being correctly performed in accordance with prescribed procedures and contract requirements.
Product Assurance [STD 8739.8]	Activities to assure that all required plans are documented, and that the plans, software products, and related documentation adhere to plans and comply with the contract and/or any memorandum of agreement/understanding.
Product Examination [NPR 8735.2]	Physical inspection, measurement, or test to ensure product conformity to prescribed technical/contract requirements.
Program [STD 8729.1]	An activity within an Enterprise having defined goals, objectives, requirements, funding, and consisting of one or more projects, reporting to the NASA Program Management Council (PMC), unless delegated to a Governing Program Management Council (GPMC).
program [STD 8719.24 Annex]	the coordinated group of tasks associated with the concept, design, manufacture, preparation, checkout, and launch of a launch vehicle and/or payload to or from, or otherwise supported by the Eastern or Western ranges and the associated ground support equipment and facilities.
Program/Project Quality Assurance Surveillance Plan (PQASP) [NPR 8735.2]	A consolidated set of detailed instructions for the performance of Government contract quality assurance actions related to a specific program/project. Examples of PQASP contents include lists of contractor documents, data, and records to be reviewed; products and product attributes to be examined; processes and process attributes to be witnessed; quality system elements/attributes to be evaluated; sampling plans; and requirements related to quality data analysis, nonconformance reporting and corrective action tracking/resolution, and final product acceptance.
Program/Project/Center [STD 8709.20]	The NASA management function for the activity (from NASA-STD 0005).

Term [Citing Document(s)]	Definitions
Programmable Logic Devices (PLD) or Complex Electronics (CE) [STD 8719.13]	A programmable logic device or PLD is an electronic component used to implement user-defined functions into digital circuits. Unlike a logic gate, which has a fixed function, a PLD has an undefined function at the time of manufacture. PLD functionality is typically described using a hardware description language (HDL), such as Verilog or VHDL, which is then converted, using PLD vendor-specific tools, into the hardware gate structure of the PLD to implement the function. PLDs can be implemented one-time or can be reconfigurable. While PLD functionality is generally described using an HDL, it can also be written in specialized variations of other programming languages. The functionality can range from a simple set of gates to a very complex set of gates (i.e. an embedded processor). The compilation of gates is hardware (regardless of the complexity) including the development of the embedded processor. For the purposes of this Standard, any software to be executed on a processor embedded within a PLD will be evaluated from a software safety perspective. The design and resulting hardware will be evaluated from a system safety perspective and is not the purview of this standard. The software tools used to generate safety critical PLDs configuration files will be evaluated from a limited COTS safety perspective as per sections 6.5 and 6.6 of this standard.
Programs [NPR 8715.3]	For the purposes of this NPR the term "programs" shall be interpreted to include programs, projects, and acquisitions.
Programs [NPR 8715.7]	For the purposes of this NPR, the term "programs" includes programs, projects, and acquisitions.
Project [STD 8729.1]	An activity designated by a program and characterized as having defined goals, objectives, requirements, Life Cycle Costs, a beginning, and an end.
Project (as used in this document) [STD 8709.20]	The organization accountable for the work being performed.
Projected Obsolete Part [STD 8739.10]	A part for which a manufacturer has issued a Product Discontinuance Notification (PDN) or other notification stating that the part will no longer be manufactured after some future date.
Projects [NPR 8715.7]	For the purposes of this NPR, the term "projects" means an ELV payload mission having defined requirements, a life cycle, a beginning, and an end. A project also has a management structure and may interface with other projects, agencies, non-Government entities, and international partners. A project yields new or revised products that directly address NASA's strategic needs.
Prompt injury [STD 8719.14]	A medical condition received as a result of the falling debris which requires (or should have required) professional medical attention within 48 hours of the impact.
proof factor [STD 8719.24 Annex]	a multiplying factor applied to the limit load or maximum expected operating environment to obtain proof load or proof pressure for use in the acceptance testing.
Proof Load [STD 8719.9]	The specific load or weight applied in performance of a proof load test (typically greater than the rated load of the LDE).

Term [Citing Document(s)]	Definitions
Proof Load Test [STD 8719.9]	A load test performed prior to first use, after major modification of the load path, or at other prescribed times. This test verifies material strength, construction, and workmanship and typically uses a load greater than the rated load.
proof pressure [STD 8719.24 Annex]	(1) the product of maximum expected operating environment and a proof factor accounting for the difference in material properties between test and service environment (such as temperature); used to give evidence of satisfactory workmanship and material quality; for example, demonstrating that the component and/or system will not deform, leak or fail; (2) may be used to establish maximum initial flaw sizes for safe-life demonstration.
Propagation [STD 8719.12]	Communication of an explosion (detonation or deflagration) from one potential explosion site to another by fire, fragment, or blast (shock wave) where the interval between explosions is long enough to limit the total overpressure at any given time to that which each explosion produces independently.
Propellant [STD 8719.12]	A solid, liquid, or hybrid chemical substance used in the production of energy or pressurized gas that is subsequently used to create movement of a fluid or to generate propulsion of a vehicle, projectile, or other object.
propellant storage tank [STD 8719.24 Annex]	any container of propellants greater than one gallon. Application of the requirements of this document to storage tanks will normally vary with the size of the tank and associated hazards. Containers less than one gallon will also be subject to operational controls, as appropriate, as would any container of flammable liquid.
Property [NPR 8715.5]	In the context of this NPR, the term "property" is intended in the broadest sense. Property includes, but is not limited to, public or privately owned land/real estate, homes, factories, livestock, natural resources, facilities, equipment, and other assets (including those on or off a range or launch or landing site). Local authorities and Projects are responsible for identifying property that requires protection. In general, the range flight safety function to protect property does not include protection of the vehicle or payload being flown in a range flight operation.
Property [STD 8719.25]	In the context of this standard, the term property is intended in the broadest sense. Property includes, but is not limited to public or privately owned land/real estate, homes, factories, livestock, natural resources, facilities, equipment, and other assets (including those on or off a range or launch or landing site). Local authorities and Programs are responsible for identifying property that requires protection per NPR 8715.5. In general, the range safety function to protect property does not include protection of the vehicle or payload being flown in a range flight operation.
Property Damage [NPR 8621.1]	Damage to any type of Government or civilian property including, but not limited to, flight hardware and software, facilities, ground support equipment, and test equipment.
Protected Noncombustible [STD 8719.11]	Noncombustible structures enclosed with partitions having a minimum of 1 hour fire-resistance rating.
Provider [STD 8719.13]	The entities or individuals that design, develop, implement, test, operate, and maintain the software products. A provider may be a contractor, a university, a separate organization within NASA, or within the same organization as the acquirer.

Term [Citing Document(s)]	Definitions
Provider [STD 8739.9] [STD 8739.8]	The entities or individuals that design, develop, implement, test, operate, and maintain the software products. A provider may be a contractor, a university, a separate organization within NASA, or within the same organization as the acquirer. The term "provider" is equivalent to "supplier" in ISO/IEC 12207, Software life cycle processes.
Provider [NPR 8000.4]	A Provider is a NASA or contractor organization that is tasked by an accountable organization (i.e., the Acquirer) to produce a product or service.
Proximate Cause [NPR 8621.1]	The event that occurred, including any conditions existing immediately before the undesired outcome, directly resulted in its occurrence, and if eliminated or modified, would have prevented it. Also, known as direct cause.
Proximity Operations [NPR 8705.2]	Two or more vehicles operating in space near enough to each other so as to have the potential to affect each other. This includes rendezvous and docking (including hatch opening), undocking, and separation (including hatch closing).
Public [NPR 8705.2]	All humans not participating in the spaceflight activity who could be potentially affected by the function or malfunction of the space system.
Public [NPR 8715.5] [STD 8719.25]	For the purposes of range safety risk management, public refers to visitors and personnel (excluding NASA workforce) inside and outside NASA-controlled locations who may be on land, on waterborne vessels, or in aircraft.
public [STD 8719.24 Annex]	all persons not in the launch essential personnel category; see also neighboring operations personnel and general public.
Public Highway [STD 8719.12]	Any street, road, or highway not under NASA custody used by the general public for any type of vehicular travel.
public safety [STD 8719.24 Annex]	safety involving risks to the general public of the US or foreign countries and/or their property (both onand off-base); includes the safety of people and property that are not involved in supporting a launch along with those that may be within the boundary of a launch site.
Public Traffic Route (PTR) [STD 8719.12]	Any public street, road, highway, navigable stream, or passenger railroad. This includes roads on NASA Centers that are open to non-essential personnel or the public for thoroughfare.
Pyrotechnic Device [STD 8719.12]	All devices and assemblies containing or actuated by propellants or explosives, with the exception of large rocket motors. Pyrotechnic devices include items such as initiators, ignitors, detonators, safe-and-arm devices, booster cartridges, pressure cartridges, separation bolts and nuts, pin pullers, linear separation systems, shaped charges, explosive guillotines, pyrovalves, detonation transfer assemblies (mild detonating fuse, confined detonating cord, confined detonating fuse, shielded mild detonating cord, etc.), thru-bulkhead initiators, mortars, thrusters, explosive circuit interrupters, and other similar items. These may be electrically (Electro-Explosive Device, EED) or mechanically initiated. (Note: OSHA defines pyrotechnics as any combustible or explosive compositions or manufactured articles designed and prepared for the purpose of producing audible or visible effects which are commonly referred to as fireworks.)

Term [Citing Document(s)]	Definitions
Qualification [STD 8739.10]	Tests consisting of mechanical, electrical, and environmental intended to verify that materials, design, performance, and long-term reliability of the part are consistent with the specification and intended application, and to assure that manufacturer processes are consistent from lot to lot.
qualification tests [STD 8719.24 Annex]	the required tests conducted under specified conditions, by, or on behalf of the government, using items representative of the production configuration in order to determine compliance with item design requirements as a basis for production approval.
Qualified Manufacturers List (QML) [STD 8739.10]	A classification issued by a qualifying agency that identifies manufacturers (along with other information) that have met certain standards for qualification.
Qualified Parts List (QPL) [STD 8739.10]	A classification issued by a qualifying agency that identifies products (along with other information) that have met certain standards for qualification.
Qualified Person [STD 8719.9]	A person who, by possession of a recognized degree in an applicable field or certificate of professional standing, or who, by extensive knowledge, training, and experience, has successfully demonstrated the ability to solve or resolve problems relating to the subject matter and work.
Quality [NPR 8715.3]	The composite of material attributes including performance features and characteristics of a product or service to satisfy a given need.
Quality Assurance Letter of Delegation (LOD) [NPR 8735.2]	Documented instructions from NASA to a Federal Agency detailing quality assurance support responsibility and services required in support of a designated contract.
Quality Assurance Support Contractor [NPR 8735.2]	A non-Government entity on contract with NASA and independent of the contractor providing supplies or services that is tasked to perform specified quality assurance surveillance functions (e.g., GMIPs).
Quality Audit, Assessment, and Review [NPR 8705.6]	An independent verification that each NASA Center, program, and project is in compliance with the applicable NASA SMA quality assurance and software assurance requirements.
Quality Conformance Inspection [STD 8739.10]	Inspection or test, used to verify conformance with requirements.
Quality Escape [NPR 8735.1]	Any product released by an internal or external supplier or sub-tier supplier that is subsequently determined to be nonconforming to contract and/or product specification requirements.
Quantity Distance (QD) [STD 8719.12]	Quantity of explosives material and distance separation relationships which provide defined types of protection.

Term [Citing Document(s)]	Definitions
Radiation Hardened (EEE Parts) [STD 8739.10]	EEE components designed to operate in man-made or natural space radiation environments and show complete immunity up to a designated level of total ionizing dose (TID) and immunity to one or more classes of single event effects (SEE). Note that standard radiation hardness assurance (RHA) designators are available on many MIL-STD marked parts; however, it is important to note that the designator may not include important areas of performance such as SEE or enhanced low dose rate susceptibility (ELDRS).
radiation source [STD 8719.24 Annex]	materials, equipment, or devices that generate or are capable of generating ionizing radiation including naturally occurring radioactive materials, by-product, source materials, special nuclear materials, fission products, materials containing induced or deposited radioactivity, nuclear reactors, radiographic and fluoroscopic equipment, particle generators and accelerators, radio frequency generators such as certain klystrons and magnetrons that produce X-rays, and high voltage devices that produce X-rays.
Radio Frequency [STD 8739.4]	The frequency spectrum from 15 kHz to 100 GHz. Cables are seldom used above 18 GHz.
Radio Frequency Interference [STD 8739.4]	Electromagnetic radiation in the radio frequency spectrum from 15 kHz to 100 GHz.
radio frequency silence [STD 8719.24 Annex]	a period of time where radio frequency (RF) transmitters/emitters, either fixed-in-place or transient, are prohibited from emitting RF energy in a specified area. It is acceptable for approved RF transmitter/emitter to be located outside of this zone and emit RF energy.
radioactive equipment or device [STD 8719.24 Annex]	equipment or devices that generate, or are capable of generating, ionizing radiation including radiographic and fluoroscopic equipment, particle generators and accelerators, radio frequency generators such as certain klystrons and magnetrons that produce X-rays, and high voltage devices that produce X-rays.
radioactive material [STD 8719.24 Annex]	materials that generate, or are capable of generating, ionizing radiation including naturally occurring radioactive materials, by-product materials, source materials, special nuclear materials, fission products, materials containing induced or deposited radioactivity, and nuclear reactors.
Radiological Control Center (RADCC) [NPR 8715.3]	A temporary information clearinghouse established on an as-needed basis to coordinate actions that could be required for mitigation, response, and recovery of an incident involving the launching of nuclear material.
Range [NPR 8621.1] [NPR 8715.5]	A permanent or temporary area or volume of land, sea, or airspace within or over which orbital, suborbital, or atmospheric vehicles are tested or flown. This includes the operation of launch vehicles from a launch site to orbital insertion or final landing or impact of suborbital vehicle components. This also includes the entry of space vehicles from the point that the commit to deorbit is initiated to the point of intact vehicle impact or landing or the impact of all associated debris. This includes range operations with aeronautical vehicles from takeoff to landing.

Term [Citing Document(s)]	Definitions
Range [STD 8719.25]	A permanent or temporary area or volume of land, sea, or airspace within or over which orbital, suborbital, or atmospheric vehicles are tested or flown. This includes the operation of launch vehicles from a launch site to orbital insertion or final landing or impact of suborbital vehicle components. This also includes the entry of space vehicles from the point where the vehicle falls below 100km to the point of intact vehicle impact or landing or the impact of all associated debris. This includes range flight operations with aeronautical vehicles from takeoff to landing.
Range Flight Operation [NPR 8715.5]	The flight of a launch or entry vehicle or experimental aeronautical vehicle including any payload, at, to, or from a range, launch site, or landing site. Range flight operations utilize specific infrastructure as well as trained and certified personnel to monitor, command, and control the range flight safety elements associated with Projects. Range flight operations do not include the flight of conventional piloted aircraft unless specific aspects of the operation require range flight safety involvement to protect the public, NASA workforce, and property. Range flight operations do not include on-orbit operations of vehicles after orbital insertion or prior to initiation of entry.
Range Flight Operation [STD 8719.25]	The flight of a launch or entry vehicle or experimental aeronautical vehicle including any payload, at, to, or from a range, launch site, or landing site. Range flight operations utilize specific infrastructure as well as trained and certified human interfaces to monitor, command, and control the range safety elements associated with Programs. Range flight operations do not include the flight of conventional piloted aircraft unless specific aspects of the operation require range safety involvement to protect the public, workforce, or property. Range flight operations do not include on orbit operations of vehicles after orbital insertion or prior to initiation of entry.
Range Flight Safety Program [NPR 8715.5]	A Program implemented to ensure that the risk to the public, NASA workforce, and property during range flight operations is effectively managed.
Range Flight Safety Program [STD 8719.25]	A Program implemented to ensure that the risk to the public, workforce, and property during range flight operations is effectively managed.
Range Operator [STD 8719.25]	A range operator is either a NASA, DoD, commercial, or foreign entity responsible for providing the ground, sea, air, or space-based assets required to support range flight operations.
range or ranges [STD 8719.24 Annex]	in this publication, range or ranges refers to the Eastern Range at Cape Canaveral Air Force Station, Kennedy Space Center, and Patrick Air Force Base, and the Western Range at Vandenberg Air Force Base.
Range Safety [NPR 8715.3]	Application of safety policies, principles, and techniques to ensure the control and containment of flight vehicles to preclude an impact of the vehicle or its pieces outside of predetermined boundaries from an abort which could endanger life or cause property damage. Where the launch range has jurisdiction, prelaunch preparation is included as a safety responsibility.

Term [Citing Document(s)]	Definitions
Range Safety [NPR 8715.5]	Application of safety policies, principles, and techniques to protect the public, NASA workforce, and/or property from hazards associated with range flight operations. Additionally, the term "Range Safety" is informally used to refer to the organization responsible for implementing/enforcing range safety requirements.
Range Safety [NPR 8715.7]	Application of safety policies, principles, and techniques to ensure the control and containment of flight vehicles to preclude an impact of the vehicle or its pieces outside of predetermined boundaries from an abort which could endanger life or cause property damage. Where the launch range has jurisdiction, prelaunch preparation is included as a safety responsibility. Additionally, the term "Range Safety" is informally used to refer to the organization responsible for implementing/enforcing range safety requirements (e.g., USAF 30th or 45th Space Wings' Safety Offices and the Wallops Flight Facility Safety Office).
Range Safety [STD 8719.25]	Application of safety policies, principles, and techniques to protect the public, workforce, or property from hazards associated with range flight operations.
Range Safety Officer (RSO) [NPR 8715.5]	A person responsible for safety during a range flight operation. An RSO has the authority to hold or abort the operation, or take a risk mitigation action, which includes terminating the flight. RSO is synonymous with the term Mission Flight Control Officer used at some DoD ranges.
Range Safety Officer (RSO) [STD 8719.25]	A person responsible for real-time safety during a range flight operation. An RSO has the authority to hold or abort the operation, or take a risk mitigation action, which includes terminating the flight. RSO is synonymous with the term MFCO used at some DoD ranges.
Range Safety Organization [STD 8719.25]	An organization that reports to the safety authority for range flight operations, oversees the implementation of range safety requirements, and may provide range safety-related services and operational support to Programs.
Range Safety Program [STD 8719.24 Annex]	a program implemented to ensure that launch and flight of launch vehicles and payloads present no greater risk to the general public than that imposed by the over-flight of conventional aircraft; such a program also includes launch complex and launch area safety and protection of national resources.
Range Safety Representative [STD 8719.24 Annex]	a government employee or member of the US Air Force assigned to the 30/45 Space Wing/Wing Safety office or a contractor employee designated and authorized by 30/45 Space Wing/Wing Safety to act on behalf of the organization.
Range Safety Waiver [STD 8719.25]	A written authorization allowing a range flight operation to continue even though a specific range safety requirement is not satisfied and the Program is not able to demonstrate an equivalent level of safety. A Range Safety Waiver involves the formal acceptance of increased safety risk by appropriate authorities.
Range User [STD 8719.25]	A range user is considered a flight test or launch or entry vehicle Program that conducts range flight operations on a range.
Rated Capacity [STD 8719.9]	See Rated Load.

Term [Citing Document(s)]	Definitions
Rated Load [STD 8719.9]	The maximum load a lifting device or equipment is designed to lift under normal operating conditions. This value may be marked on the device indicating maximum capacity. This is also the load referred to as "safe working load or the working load limit." If the device has never been downrated or uprated, this also is the "manufacturer rated load."
rated load [STD 8719.24 Annex]	An assigned weight that is the maximum load the device or equipment shall operationally handle and maintain. This value is marked on the device indicating maximum working capacity. This is also the load referred to as "safe working load" or "working load limit." If the device has never been downrated or uprated, this also is the "manufacturer's rated load."
Reader [STD 8739.9]	An inspection participant who describes the product being inspected to the other inspectors during the inspection meeting. (see Wiegers 2008)
Real-time Waiver Request [STD 8709.20]	A request for relief that was unforeseen and is needed on a short term basis to meet an immediate operational need.
Recertification [STD 8719.17]	The renewal of a previous certification with adjustments as necessary to accommodate new information, configuration or operating parameter changes, or PVS degradation.
Recertification File [STD 8719.24 Annex]	a file that contains data showing that a specific piece of material handling equipment/material ground support equipment meets the periodic test and inspection requirements of this document.
Recognized Licensing Organization [STD 8719.9]	Operator licensing organization meeting industry recognized criteria for written testing materials, practical examinations, test administration, grading, facilities/equipment, and personnel.
Recommendation [NPR 8621.1]	An action developed by the investigating authority to correct the cause or a finding identified during the investigation.
Record Review [NPR 8735.2]	Review and verification that recorded data properly evidences conformance to contract requirements (e.g., invoked drawings, specifications). Recorded data may document work performance, product attributes, product configuration, product performance, or quality assurance actions performed by the contractor.
Recorder [STD 8739.9]	An inspection participant who documents the defects and issues brought up during the inspection meeting. (see Wiegers 2008)
Recovery System [STD 8719.25]	A system that is installed on a flight test, launch, or entry vehicle that may be activated as planned (e.g. to preserve hardware for possible reuse) or when the vehicle has malfunctioned and cannot be recovered under its own capacity. Recovery systems are intended to preserve the vehicle and do not necessarily address range safety concerns.
Red Plague (Cu2O) [STD 8739.10]	The sacrificial corrosion of copper in a galvanic interface between silver and copper, resulting in the formation of red cuprous oxide (Cu2O). Continued exposure to an oxygen rich environment can then lead to black cupric oxide (CuO). Galvanic corrosion is promoted by the presence of moisture and oxygen at an exposed copper-silver interface (i.e., conductor end, pinhole, scratch, nick, etc.).

Term [Citing Document(s)]	Definitions
Redundancy [NPR 8715.3] [STD 8729.1]	Use of more than one independent means to accomplish a given function.
Redundancy (of design) [STD 8729.1]	A design feature which provides a system with more than one function for accomplishing a given task so that more than one function must fail before the system fails to perform the task. Design redundancy requires that a failure in one function does not impair the system's ability to transfer to a second function.
redundant [STD 8719.24 Annex]	a situation in which two or more independent means exist to perform a function.
referee fluid [STD 8719.24 Annex]	a compatible fluid, other than that used during normal system operations, that is used for test purposes because it is safer due to characteristics such as less (or non-) explosive, flammable, or toxic and/or easier to detect.
Referee Magnification Levels [STD 8739.6]	Higher levels of magnification than the maximum magnification limit defined in the applicable workmanship standard, used for closer examination of an anomaly to determine if it is a defect.
Refraction [STD 8739.5]	The bending of a beam of light in transmission through an interface between two dissimilar media or in a medium whose refractive index is a continuous function of position, for example graded index medium.
Regular Service LDE [STD 8719.9]	LDE used one or more times per month.
Release ID [STD 8739.9]	Identification code associated with a product's version level.
Reliability [NPR 8705.2] [NPR 8715.3] [STD 8739.9]	The probability that a system of hardware, software, and human elements will function as intended over a specified period of time under specified environmental conditions.
Reliability [STD 8729.1]	The probability that an item will perform its intended function for a specified interval under stated conditions. The function of an item may be composed of a combination of individual subfunctions to which the top-level reliability value can be apportioned.
Reliability Analyses [STD 8729.1]	A set of conceptual tools and activities used in reliability engineering. Examples of common analyses are Failure Modes and Effects Analysis (FMEA) and Fault Tree Analysis in failure space and Reliability Block Diagram Analysis (RBDA) in success space.
Reliability Analysis [NPR 8715.3]	An evaluation of reliability of a system or portion thereof. Such analysis usually employs mathematical modeling, directly applicable results of tests on system hardware, estimated reliability figures, and non-statistical engineering estimates to ensure that all known potential sources of unreliability have been evaluated.
Reliability Assurance [STD 8729.1]	The management and technical integration of the reliability activities essential in maintaining reliability performance, including design, production, risk management, and product assurance activities.

Term [Citing Document(s)]	Definitions
Reliability Block Diagram Analysis (RBDA) [STD 8729.1]	A deductive (top-down) method that generates a symbolic-logic model in success space that depicts and analyzes the reliability (and/or availability) relationships between the system and system elements and/or events. Typical RBD models are constructed of series, parallel, and/or combinations of series and parallel configurations. The RBD model describes a successful operation when an uninterrupted path exists between the model's input and output. The RBDA process, for example, provides a design baseline and serves as a means to identify weak areas and changes early in the design phase and serves as input to accomplish related analyses (e.g., FMEA, FTA, spare, and maintenance).
Reliability Centered Maintenance [STD 8729.1]	An on-going process that determines the mix of corrective and preventive maintenance practices to provide the required reliability at the minimum cost. It can use diagnostic tools and measurements to assess when a component is near failure and should be replaced. The basic thrust is to eliminate more costly corrective maintenance and minimize preventive maintenance.
Relief [STD 8709.20]	A waiver, deviation, or request for determination of non-applicability to modify or eliminate a stated requirement and usually not meet the full intent and letter of the requirement as stated.
Relief Adjudication [STD 8709.20]	See "Adjudication"
remote control [STD 8719.24 Annex]	control of a system from a remote and safe location.
Remote Emergency Stop (Remote E-Stop) [STD 8719.9]	A manually operated switch or valve to cut off electric or fluid power independently of the regular operating controls that is located remotely from the operator control station.
render safe [STD 8719.24 Annex]	an action to bring to a safe condition.
Repair [STD 8739.5]	Action on a nonconforming product to make it acceptable for the intended use.
Repair [STD 8739.4]	The action on a nonconforming product to make it acceptable for the intended use.
Request for Relief [STD 8719.9]	Documented request for permission to perform some act contrary to established requirements.
required (in reference to instrumentation or capability) [STD 8719.24 Annex]	a system that must be made operationally ready to support Range Safety.
Requirement [STD 8739.9]	A precise statement of need intended to convey understanding about a condition or capability that must be met or possessed by a system or system component to satisfy a contract, standard, specification, or other formally imposed document. The set of all requirements forms the basis for subsequent development of the system or system components.

Term [Citing Document(s)]	Definitions
Requirement Relief Request [STD 8709.20]	Request for a waiver, deviation, or determination of non-applicability.
Requirements [STD 8729.1]	Requirements are statements of need that define what a system will do and how well it must perform those tasks.
Requirements Evaluation and Documentation Assessment & Analysis [NPR 8705.6]	1) An independent document-only desk review of the flow down of SMA requirements to the NASA Center documents and to designated programs and projects documents; and 2) a top-level review of designated NASA contracts in support of IFOSA and QAAR.
Requirements Traceability [STD 8709.20]	The process of mapping originating requirements to implementing requirements. This is completed after a determination of applicable requirements is completed.
Rescue [NPR 8705.2]	The process of locating the crew, proceeding to their position, providing assistance, and transporting them to a location free from danger.
Residual Risk [NPR 8715.3]	The level of risk that remains after applicable safety-related requirements have been satisfied. In a risk-informed context, such requirements may include measures and provisions intended to reduce risk from above to below an acceptable level.
residual stress [STD 8719.24 Annex]	the stress that remains in a structure after processing, fabrication, assembly, testing, or operation; for example, welding induced residual stress.
Resin [STD 8739.1]	Generally, any synthetic organic material produced by polymerization.
resource safety [STD 8719.24 Annex]	the protection of facilities, support equipment, or other property from damage due to mishaps; also known as resource protection.
Responsible Organization [NPR 8621.1]	The organization responsible for the activity, people, operation, or program, where a mishap occurs or the lowest level of organization where corrective action will be implemented.
Responsible Organization [STD 8719.9]	Entity or a representative thereof responsible for the design, operation, maintenance, testing, inspection, or personnel training and licensing of LDE. (In some cases, this may be the LDEM).
rest period [STD 8719.24 Annex]	the period of time immediately prior to the beginning of the duty period; for launch-essential personnel, it is mandatory that the rest period include the time necessary for meals, transportation, and 8 hours of uninterrupted rest prior to reporting for duty. Rest periods in preparation for launch operations will start no earlier than 2 hours after the assigned personnel are released from an earlier launch or range operations. Only the Chief of Safety or Space Wing Commander has the authority to waive the safety rest period requirements for Mission Ready (Category A) personnel; see also crew rest.
Restricted Area [STD 8719.12]	Any area, usually fenced, at an establishment where the entrance and egress of personnel and vehicular traffic are controlled for reasons of safety.

Term [Citing Document(s)]	Definitions
Reusable Launch Vehicle [NPR 8715.5] [STD 8719.25]	Experimental or operational space launch vehicle that is intended to be reused (at least in part).
Review [STD 8739.8]	A process or meeting during which a software product or related documentation is presented to project personnel, customers, managers, software assurance personnel, users or user representatives, or other interested parties for comment or approval. [IEEE 610.12, IEEE Standard Glossary of Software Engineering Terminology] Reviews include, but are not limited to, requirements review, design review, code review, test readiness review. Other types may include peer review and formal review.
Review [NPR 8705.6]	An activity that proposes to figure out how well the thing being reviewed is capable of achieving established objectives. Reviews ask the following question: is the subject (or object) of the review a suitable, adequate, effective, and efficient way of achieving established objectives.
Review [STD 8729.1]	A critical examination of a task or program/project to determine compliance with requirements and objectives.
Rework [STD 8739.5]	Action on a nonconforming product to make it conform to the requirements.
Rework [STD 8739.4]	The action on a nonconforming product to make it conform to the requirements.
Rigging Hardware [STD 8719.9]	A detachable load supporting device such as a shackle, link, eyebolt, ring, swivel, or clevis.
Right Ascension of Ascending Node [STD 8719.14]	The angle between the line extending from the center of the Earth to the ascending node of an orbit and the line extending from the center of the Earth to the vernal equinox, measured from the vernal equinox eastward in the Earth's equatorial plane.
Risk [NPR 8000.4]	Risk is the potential for shortfalls with respect to achieving explicitly established and stated objectives. As applied to programs and projects, these objectives are translated into performance requirements, which may be related to mission execution domains (safety, mission success, cost, and schedule) or institutional support for mission execution. Risk is operationally characterized as a set of triplets: The scenario(s) leading to degraded performance with respect to one or more performance measures (e.g., scenarios leading to injury, fatality, destruction of key assets; scenarios leading to exceedance of mass limits; scenarios leading to cost overruns; scenarios leading to schedule slippage). The likelihood(s) (qualitative or quantitative) of those scenarios. The consequence(s) (qualitative or quantitative severity of the performance degradation) that would result if those scenarios were to occur. Uncertainties are included in the evaluation of likelihoods and identification of scenarios.

Term [Citing Document(s)]	Definitions
Risk [NPR 8621.1]	In the context of mission execution, risk is operationally defined as a set of triplets: a. The scenarios leading to degraded performance with respect to one or more performance measures (e.g., scenarios leading to [1] injury, fatality, destruction of key assets; [2] exceedance of mass limits; [3] cost overruns; or [4] schedule slippage). b. The likelihoods (qualitative or quantitative) of those scenarios. c. The consequences (qualitative or quantitative severity of performance degradation) that would result if those scenarios were to occur.
Risk [NPR 8705.2]	The combination of (1) the probability (qualitative or quantitative) including associated uncertainty that the space system will experience an undesired event (or sequences of events) such as internal system or component failure or an external event and (2) the magnitude of the consequences (personnel, public, and mission impacts) and associated uncertainties given that the undesired event(s) occur(s).
Risk [NPR 8715.3]	The combination of (1) the probability (qualitative or quantitative) of experiencing an undesired event, (2) the consequences, impact, or severity that would occur if the undesired event were to occur and (3) the uncertainties associated with the probability and consequences.
Risk [NPR 8715.5] [STD 8719.25]	A measure that takes into account both the probability of occurrence and the consequence of a hazard or combination of hazards to a population or installation. Unless otherwise noted, risk to people is measured in casualties and expressed as individual risk or collective risk.
Risk [STD 8719.7]	The combination of the hazard severity with the likelihood of its occurrence.
Risk [STD 8729.1]	In the context of mission execution, risk is operationally defined as a set of triplets. [1] The scenario(s) leading to degraded performance with respect to one or more performance measures (e.g., scenarios leading to injury, fatality, destruction of key assets; scenarios leading to exceedance of mass limits; scenarios leading to cost overruns; scenarios leading to schedule slippage). [2] The likelihood(s) (qualitative or quantitative) of those scenarios. [3] The consequence(s) (qualitative or quantitative severity of the performance degradation) that would result if those scenarios were to occur. Uncertainties are included in the evaluation of likelihoods and consequences.
risk [STD 8719.24 Annex]	a measure that takes into consideration both the probability of occurrence and the consequence of a hazard to a population or installation. Risk is measured in the same units as the consequence such as number of injuries, fatalities, or dollar loss. For Range Safety, risk is expressed as casualty expectation or shown in a risk profile; see also collective risk and individual risk.
Risk (Safety) Assessment [NPR 8715.3] [NPR 8715.7]	Process of qualitative risk categorization or quantitative risk (safety) estimation, followed by the evaluation of risk significance.
Risk Acceptability Criterion [NPR 8000.4]	A rule for determining whether a given organizational unit has the authority to decide to accept a risk.

Term [Citing Document(s)]	Definitions
Risk Acceptance [STD 8729.1]	The formal process of justifying and documenting a decision not to mitigate a given risk associated with achieving given objectives or given performance requirements.
risk analysis [STD 8719.24 Annex]	a study of potential risk.
Risk Assessment [NPR 8705.2]	An evaluation of a risk item that determines (1) what can go wrong, (2) how likely is it to occur, and (3) what the consequences are.
Risk Assessment Code [STD 8719.17]	A numerical expression of comparative risk of a condition determined by an evaluation of both the potential severity of a consequence and the likelihood of that consequence occurring.
Risk Management [NPR 8000.4]	Risk management includes RIDM and CRM in an integrated framework. This is done in order to foster proactive risk management, to inform better decision making through better use of risk information, and then to manage more effectively implementation risks by focusing the CRM process on the baseline performance requirements informed by the RIDM process.
Risk Management [NPR 8715.3]	An organized, systematic decision-making process that efficiently identifies, analyzes, plans, tracks, controls, communicates, and documents risk to increase the likelihood of achieving project goals.
Risk Management [STD 8729.1]	An organized, systematic decision-making process that efficiently identifies, analyzes, plans, tracks, controls, communicates, and documents risk and establishes mitigation approaches and plans to increase the likelihood of achieving program/project goals.
Risk Ranking [NPR 8705.2]	The ordering of risk contributors such as accident scenarios or classes of accident scenarios based on the extent of their contribution (accounting for hazard controls, crew survival capabilities, and other risk reduction measures) such that the significant contributors can be identified.
Risk Reduction [STD 8729.1]	The modification of a process, system, or activity in order to reduce a risk by reducing its probability, consequence severity, or uncertainty, or by shifting its timeframe.
Risk Review Boards [NPR 8000.4]	Formally established groups of people assigned specifically to review risk information. Their output is twofold: (1) to improve the management of risk in the area being reviewed and (2) to serve as an input to decision-making bodies in need of risk information.
Risk-Informed Decision Making (RIDM) [NPR 8000.4]	A risk-informed decision-making process uses a diverse set of performance measures (some of which are model-based risk metrics) along with other considerations within a deliberative process to inform decision making.
Root Cause [NPR 8621.1]	An event or condition, primarily associated with organizational factors, which existed before the intermediate cause and directly resulted in its occurrence (indirectly caused or contributed to the proximate cause and subsequent undesired outcome) and, if eliminated or modified, would have prevented the intermediate cause from occurring and the undesired outcome. Typically, multiple causes contribute to an undesired outcome. In the absence of a prevalent organizational factor, the root cause may be identified as undetermined.

Term [Citing Document(s)]	Definitions
Root Cause Analysis [NPR 8621.1]	A structured evaluation method used to identify the root causes of an undesired outcome and the actions adequate to prevent occurrence. Root cause analysis should continue until organizational factors have been identified or until data are exhausted.
safe and arm device [STD 8719.24 Annex]	a device that provides mechanical interruption (safe) or alignment (arm) of the explosive train and electrical interruption (safe) or continuity (arm) of the firing circuit.
Safe Haven [NPR 8705.2]	A functional association of capabilities and environments that is initiated and activated in the event of a potentially life-threatening anomaly and allows human survival until rescue, the event ends, or repair can be affected.
Safe Working Load [STD 8719.9]	See Rated Load.
Safety [NPR 8000.4]	In a risk-informed context, safety is an overall condition that provides sufficient assurance that mishaps will not result from the mission execution or program implementation, or, if they occur, their consequences will be mitigated. This assurance is established by means of the satisfaction of a combination of deterministic criteria and risk-informed criteria.
Safety [NPR 8705.2]	The absence from those conditions that can cause death, injury, occupational illness, damage to or loss of equipment or property, or damage to the environment.
Safety [NPR 8715.3]	Freedom from those conditions that can cause death, injury, occupational illness, damage to or loss of equipment or property, or damage to the environment. In a risk-informed context, safety is an overall mission and program condition that provides sufficient assurance that accidents will not result from the mission execution or program implementation, or, if they occur, their consequences will be mitigated. This assurance is established by means of the satisfaction of a combination of deterministic criteria and risk criteria.
Safety [NPR 8715.7]	Freedom from those conditions that can cause death, injury, occupational illness, damage to or loss of equipment or property, or damage to the environment. In a risk-informed context, safety is an overall mission and program condition that provides sufficient assurance that mishaps will not result from the mission execution or program implementation, or, if they occur, their consequences will be mitigated. This assurance is established by means of the satisfaction of a combination of deterministic criteria and risk-informed criteria.
Safety Analysis [NPR 8715.3] [NPR 8715.7]	Generic term for a family of analyses, which includes but is not limited to, preliminary hazard analysis, system (subsystem) hazard analysis, operating hazard analysis, software hazard analysis, sneak circuit, and others.
Safety Analysis Report (SAR) [NPR 8715.3]	A safety report of considerable detail prepared by or for the program detailing the safety features of a particular system or source.
Safety Analysis Summary (SAS) [NPR 8715.3]	A brief summary of safety considerations for minor sources; a safety report of less detail than the SAR.

Term [Citing Document(s)]	Definitions
Safety and Mission Assurance Requirement Tracking System – (SMARTS) [STD 8709.20]	NASA data system that contains meta data about Agency-wide SMA requirements (SMARTS can be accessed at: http://smarts.nasa.gov).
Safety Assurance [NPR 8715.3]	Providing confidence that acceptable risk for the safety of personnel, equipment, facilities, and the public during and from the performance of operations is being achieved.
safety clearance zones [STD 8719.24 Annex]	the restricted areas designated for day-to-day prelaunch processing and launch operations to protect the public, launch area, and launch complex personnel; these zones are established for each launch vehicle and payload at specific processing facilities, including launch complexes; includes hazard clearance area and hazardous launch area.
Safety Critical [NPR 8715.3] [NPR 8715.7] [STD 8719.13]	Term describing any condition, event, operation, process, equipment, or system that could cause or lead to severe injury, major damage, or mission failure if performed or built improperly, or allowed to remain uncorrected.
safety critical [STD 8719.24 Annex]	an operation, process, system, or component that controls or monitors equipment, operations, systems, or components to ensure personnel, launch area, and public safety; may be hazardous or non-hazardous.
safety critical computer system function [STD 8719.24 Annex]	a computer function containing operations that, if not performed, if performed out of sequence, or if performed incorrectly, may result in improper or lack of required control functions that may directly or indirectly cause a hazard to exist.
Safety Critical Event [STD 8729.1]	An event (successful or failure) of whose proper recognition, control, performance or tolerance is essential to safe system operation or use.
safety critical facility [STD 8719.24 Annex]	a hazardous facility or a facility that is used to store, handle, or process systems determined to be safety critical by Range Safety.
safety critical procedure [STD 8719.24 Annex]	a designation for a particular type of Range User procedure; a document containing steps in sequential order used to reliably process safety critical systems or conduct safety critical operations; non-hazardous safety critical procedures have no specific content requirements but do require Range Safety review and approval.
safety critical software [STD 8719.24 Annex]	software deemed to be safety critical per the litmus test in NASA-STD-8719.3, NASA Software Safety Standard.
Safety Culture	The value placed on safety as demonstrated by people's behavior. It is the way safety is perceived, valued and prioritized in an organization. It reflects the commitment to safety at all levels in the organization. It is "how an organization behaves when no one is watching". Safety culture is expressed and observed via individual and group attitudes and behavior, and organizational processes.

Term [Citing Document(s)]	Definitions
Safety Data Package [NPR 8715.7]	A data submittal that provides a detailed description of hazardous and safety critical flight hardware equipment, systems, components and materials that comprise the payload. Includes hazard reports, safety assessments, inhibits, and mitigations. Known as a Missile System Prelaunch Safety Package (MSPSP) with respect to AFSPCMAN 91-710, Range Safety User Requirements.
Safety Data Package [STD 8719.13]	As used within this document refers to all or a combination of Program/Project/Facility failure modes and effect analysis (FMEA), hazard analysis/report, criticality analysis, reliability analysis, computing system safety analysis, orbital debris assessment report/end of mission plan, and documentation utilized to objectively demonstrate the software measures taken to control/mitigate hazards/conditions/events have been verified and successfully implemented. Other analysis such as the software requirements, software design analysis, mapping between the software safety requirements and the safety data package can be included in the Software Safety Case Study.
Safety Device [NPR 8715.3]	A device that is part of a system, subsystem, or equipment that will reduce or make controllable hazards which cannot be otherwise eliminated through design selection.
Safety Evaluation Report (SER) [NPR 8715.3]	A safety report prepared by the INSRP detailing the INSRP's assessment of the nuclear safety of a particular source or system based upon INSRP's evaluation of the program-supplied SAR and other pertinent data.
Safety Factor [STD 8719.9]	See Design Factor.
safety factor [STD 8719.24 Annex]	for pressure systems, the ratio of design burst pressure over the maximum allowable working pressure or as design pressure; for mechanical systems, it can also be expressed as the ratio of tensile or yield strength over the maximum allowable stress of the material.
Safety Goal [NPR 8705.2]	The level of safety that serves as a long-term target for repeatedly flown missions, specified at the system level in terms of an aggregate measure of risk to the crew such as the probability of a loss of crew.
safety holds [STD 8719.24 Annex]	the holdfire capability, emergency voice procedures, or light indication system of each launch system used to prevent launches in the event of loss of Range Safety critical systems or violations of mandatory Range Safety launch commit criteria.
Safety Margin [NPR 8715.3]	Difference between as-built factor of safety and the ratio of actual operating conditions to the maximum operating conditions specified during design.
Safety Oversight [NPR 8715.3]	Maintaining functional awareness of program activities on a real-time basis to ensure risk acceptability.
Safety Program [NPR 8715.3] [NPR 8715.7]	The implementation of a formal comprehensive set of safety procedures, tasks, and activities to meet safety requirements, goals, and objectives.
Safety Threshold [NPR 8705.2]	The minimum tolerable level of safety for a given reference mission, specified at the system level in terms of an aggregate measure of risk to the crew such as the probability of a loss of crew.

Term [Citing Document(s)]	Definitions
Safety-Critical GMIP [NPR 8735.2]	GMIPs performed to ensure compliance with contract requirements that, if violated, can credibly result in loss of human life. This includes witness or verification of hardware, manufacture, assembly, integration, test, maintenance, operation, or nonconformance resolution tasks which, if incorrectly accomplished, could result in loss of life.
Safety-Critical Product [NPR 8735.1]	A product whose failure, malfunction, or absence could cause a catastrophic or safety-critical failure resulting in the loss of or serious damage to the craft/vehicle/facility, loss of mission, an unacceptable risk of personal injury, or loss of life.
safing procedures [STD 8719.24 Annex]	the process of taking a system that is in a hazardous configuration and performing those tasks necessary to bring it to a condition which is safe for further activities; safing procedures are part of the backout procedures for a system.
Scenario [NPR 8000.4]	A sequence of events, such as an account or synopsis of a projected course of action or events.
Screening [STD 6008]	An in-house receiving inspection that verifies that requested procurement documentation has been received and that procurement requirements have been met. This is intended to be done in addition to the CVT required.
Screening [STD 8739.10]	Tests, typically applied to 100% of parts in a lot, intended to remove nonconforming parts (parts with random defects that are at increased risk of resulting in early failures, known as infant mortality) from an otherwise acceptable lot and thus increase confidence in the reliability of the parts selected for use.
Sealing Plug [STD 8739.4]	A plug that is inserted to fill an unoccupied contact aperture in a connector. Its function is to seal an unoccupied aperture in the assembly, especially in environmental connectors.
Secondary Payload [NPR 8715.7]	Often payloads are launched with excess performance capability due to large spacecraft being "volume limited" rather than "mass limited." To fully utilize this excess capability, secondary payloads can be launched along with the primary, providing a means to economically launch small spacecraft. These are generally independent missions that minimally impact the primary payload.
Secure Flight Termination System [STD 8719.25]	National Security Agency approved cryptography incorporated into the operations center and vehicle that provides a capability for the secure or authenticated transmissions of a flight termination command or the activation of the FTS.
Semi-major axis [STD 8719.14]	Half the sum of the distances of apogee and perigee from the center of the Earth (or other body) equal to half the length of the major axis of the elliptical orbit.
Senior Fire Officer [STD 8719.11]	A fire department's Fire Chief or his/her designee.
separate power source [STD 8719.24 Annex]	a dedicated and independent source of power.
Serious [NPR 8715.3]	When used with "hazard," "violation," or "condition," denotes there is a substantial probability that death or serious physical harm could result.

Term [Citing Document(s)]	Definitions
Serious Workplace Hazard [NPR 8621.1]	A condition, practice, method, operation, or process having substantial probability of death or serious physical harm.
service life [STD 8719.24 Annex]	(1) the total life expectancy of a part or structure; service life starts with the manufacture of the structure and continues through all acceptance testing, handling, storage, transportation, operations, refurbishment, retesting, and retirement; (2) the period of time between the initial lot acceptance testing and the subsequent age surveillance testing for ordnance.
Service Magazine [STD 8719.12]	A building used for the intermediate storage of explosives materials not exceeding the minimum amount necessary for safe efficient production.
Severity [STD 8739.9]	A degree or category of magnitude for the ultimate impact or effect of executing a given software fault, regardless of probability.
Severity (of a failure) [STD 8729.1]	A measure of the effect or consequence of a failure in relation to mission performance, hazards to material or personnel, and maintenance cost. Programs/projects typically establish their own severity definitions and classifications.
Shall [STD 8709.20]	The verb "shall" indicates a mandatory requirement. The collaborative CMO tasks use the emphatic "shall" to indicate an obligation or requirement on the part of the Supplier. (An exception is when the Supplier and Program/Project/Center and/or CMO are involved in a collaborative requirement. The emphatic "shall" is used as a grammatical convenience and does not imply special expectation on the Program/Project/Center. Example: The Program/Project/Center and the Supplier shall perform….) (from NASA-STD 0005).
Shall [STD 8739.1]	Signifies a requirement statement herein.
Shall [STD 8719.11]	The word "shall" indicates that the requirement is mandatory and must be followed.
shall [STD 8719.24 Annex]	as used in requirements documents, denotes a mandatory action.
shield (RF) [STD 8719.24 Annex]	a metallic barrier that completely encloses a device for the purpose of preventing or reducing induced energy.
Shielded Cable [STD 8739.4]	Cable surrounded by a metallic covering intended to minimize the effects of electrical crosstalk interference or signal radiation.
Shielding [STD 8739.4]	The metal covering surrounding one or more conductors in a circuit to prevent interference or signal radiation.
Should [STD 8709.20]	Good practices, guidance, or options are specified with the nonemphatic verbs "should," "may," or "can" (from NASA-STD 0005).
Should [STD 8739.1]	Signifies statement of a recommended practice herein.

Term [Citing Document(s)]	Definitions
Should [STD 8719.11]	The word "should" indicates that the rule is a recommendation, the advisability of which depends on the facts in each situation.
should [STD 8719.24 Annex]	as used in requirements documents, denotes a good practice and is recommended, but not required.
Significant Problem [NPR 8735.1]	Any problem that is of the highest category of significance by virtue of the problem's impact on personnel safety or mission accomplishment (schedule and objectives).
Single Event Effect (SEE) [STD 8739.10]	A generalized category of anomalies that result from a single ionizing particle. This term includes such effects as single event upsets, transients, latch–up, permanent upset, and device burnout.
Single Event Upset (SEU) [STD 8739.10]	An unintentional change in the state of a digital device, resulting in erroneous data or control induced by ionizing radiation. The change of a state is not permanent in that complete functionality can be restored by reprogramming.
Single Failure Point [NPR 8715.3]	An independent element of a system (hardware, software, or human) the failure of which would result in loss of objectives, hardware, or crew.
Single Failure Point [STD 8719.9]	A single item or component whose failure would cause an undesired event such as dropping a load or loss of control.
single failure point [STD 8719.24 Annex]	an independent element of a system (hardware, software, or human) the failure of which would result in loss of objectives, hardware, or crew; in general, a component that, if failed, could lead to the overall failure of the system (for example, in a mechanical system, a component such as a lug, link, shackle, pin, bolt, rivet, or a weld that, if failed, could cause a system inability to support a load using load path analysis).
Single Point Failure [STD 8729.1]	An independent element of a system (hardware, software, or human), the failure of which would result in loss of objectives, hardware, or crew.
single point ground [STD 8719.24 Annex]	the one interconnection for a grounded circuit with other circuits.
single point of contact [STD 8719.24 Annex]	the Range User's one point of contact for range operations.
Single-Mode Fiber [STD 8739.5]	An optical fiber in which only the lowest order bound mode can propagate at the wavelength of interest.
Sling [STD 8719.9]	A flexible lifting assembly and incorporated hardware used between a lifting device and the payload being lifted. Common types include wire rope slings, synthetic roundslings, metal mesh slings, synthetic web slings, and chain slings.
sling [STD 8719.24 Annex]	a lifting assembly and associated hardware used between the actual object being lifted (load) and the hoisting device hook.

Term [Citing Document(s)]	Definitions
Small Arms Ammunition [STD 8719.12]	Ammunition comprising a complete round, cartridge or its components, including bullets or projectiles, cartridge cases, primers/caps and propellants that are used in small arms not exceeding 12.7 mm (50 caliber or 0.5 inch) for rifle or pistol cartridges or 8 gauge for shotguns.
SMA-Sponsored Level B Training Center [STD 8739.6]	A workmanship training center that has been identified and sponsored by the local NASA Center's SMA authority to train contractors and civil servants who work inside or outside of the Center. An SMA-Sponsored Training Center may consist of more than one location where Level B trainers provide instruction and one or more instructors. See A.2.2.
Smoke Removal System [STD 8719.11]	An interconnected system of fans, ducts, dampers, and automatic and manual controls designed to effectively remove smoke and other products of combustion from select facility areas. Its use is primarily intended to compensate for the lack of a readily available means to ventilate buildings during and after structural fires, such as in below-grade or windowless building areas.
soft goods [STD 8719.24 Annex]	the nonmetal materials in a pressure system that are used to form a seal or seat for metal-to-metal contact or between other hard surfaces.
Software [STD 8719.13]	Computer programs, procedures, rules, and associated documentation and data pertaining to the development and operation of a computer system. Software also includes COTS, GOTS, MOTS, embedded software, reuse, heritage, legacy, auto generated code, firmware, and open source software components. (Definition from source document: NPD 7120.4, NASA Engineering and Program/Project Management) For the purposes of this standard, the term software will also include scripts, glueware, and wrappers.
Software [STD 8739.8]	Computer programs, procedures, rules, and associated documentation and data pertaining to the development and operation of a computer system. Software includes programs and operational data contained in hardware (e.g., firmware, programmable logic, and programmable gate arrays). This also includes COTS, GOTS, MOTS, reuse, legacy, and heritage software products and components.
Software [STD 8739.9]	Computer programs, procedures, scripts, rules, and associated documentation and data pertaining to the development and operation of a NASA component or computer system. Software includes programs and data. This also includes COTS, GOTS, MOTS, reused software, auto generated code, embedded software, firmware, the software which runs on programmable logic devices (PLDs) operating systems, and open source software components.
Software Assurance [STD 8719.13]	The planned and systematic set of activities that ensure that software life-cycle process and products conform to requirements, standards, and procedures. For NASA this includes the disciplines of software quality (functions of software quality engineering, software quality assurance, and software quality control), software safety, software reliability, software verification and validation, and IV&V.

Term [Citing Document(s)]	Definitions
Software Assurance [STD 8739.8]	The planned and systematic set of activities that ensure that software life cycle processes and products conform to requirements, standards, and procedures. [IEEE 610.12, IEEE Standard Glossary of Software Engineering Terminology] For NASA this includes the disciplines of Software Quality (functions of Software Quality Engineering, Software Quality Assurance, Software Quality Control), Software Safety, Software Reliability, Software Verification and Validation, and IV&V.
Software Assurance [STD 8739.9]	The planned and systematic set of activities that ensure that software life cycle processes and products conform to requirements, standards, and procedures. For NASA this includes the disciplines of Software Quality (functions of Software Quality Engineering, Software Quality Assurance, Software Quality Control), Software Safety, Software Reliability, Mission Software Cybersecurity Assurance, Software Verification and Validation, and IV&V.
Software Assurance Program Metrics [STD 8739.8]	Metrics related to the activities defined in the Software Assurance Program. Examples include number of reviews/audits planned vs. reviews/audits performed, software assurance effort planned vs. software assurance effort actual, and corrective actions opened vs. corrective actions closed.
Software Assurance Record [STD 8739.8]	A record that provides objective evidence of the extent of the fulfillment of the requirements for software quality, safety, reliability, verification and validation, and, when present, IV&V. This includes documentation of the software assurance activities and analyses results.
software design description [STD 8719.24 Annex]	a representation of a software system created to facilitate analysis, planning, implementation, and decision-making; a blueprint or model of the software system; used as the primary medium for communicating software design information.
Software Engineering [STD 8739.9]	The application of a systematic, disciplined, quantifiable approach to the development, operation, and maintenance of software: that is, the application of engineering to software.
Software Hazard Analysis [NPR 8715.3]	Identification and verification of adequate software controls and inhibits; and the identification, analysis, and elimination of discrepancies relating to safety critical command and control functions.
Software Life Cycle [STD 8739.8]	The period of time that begins when a software product is conceived and ends when the software is no longer available for use. The software life cycle typically includes a concept phase, requirements phase, design phase, implementation phase, test phase, installation and checkout phase, operation and maintenance phase, and sometimes, retirement phase. [IEEE 610.12, IEEE Standard Glossary of Software Engineering Terminology]
Software Life Cycle [STD 8739.9]	The period of time that begins when a software product is conceived and ends when the software is no longer available for use. The software life cycle (LC) typically includes a concept phase, requirements phase, design phase, implementation phase, test phase, installation and checkout phase, operation and maintenance phase, and sometimes, retirement phase. There are many development LC possibilities including waterfall, agile, model based, and others, all development LCs can reside within the total software lifecycle. The SW LC, usually resides within a system LC and may be the same as the system LC or can be standalone since SW can be updated when the system may not be able to be updated.

Term [Citing Document(s)]	Definitions
Software Product [STD 8739.9]	Software products include requirements, design, code, plans, user manual, etc.
Software Product Quality [STD 8739.8]	A measure of software that combines the characteristics of low defect rates and high user satisfaction.
Software Quality [STD 8739.8]	The discipline of software quality is a planned and systematic set of activities to ensure quality is built into the software. It consists of software quality assurance, software quality control, and software quality engineering. As an attribute, software quality is (1) the degree to which a system, component, or process meets specified requirements; or (2) the degree to which a system, component, or process meets customer or user needs or expectations. [IEEE 610.12, IEEE Standard Glossary of Software Engineering Terminology]
Software Quality Assurance [STD 8739.8]	The function of software quality that assures that the standards, processes, and procedures are appropriate for the project and are correctly implemented.
Software Quality Control [STD 8739.8]	The function of software quality that checks that the project follows its standards, processes, and procedures, and that the project produces the required internal and external (deliverable) products.
Software Quality Engineering [STD 8739.8]	The function of software quality that assures that quality is built into the software by performing analyses, trade studies, and investigations on the requirements, design, code, and verification processes and results to assure that reliability, maintainability, and other quality factors are met.
Software Quality Metrics [STD 8739.8]	Metrics are quantitative values that measure the quality of software or the processes used to develop the software, or some attribute of the software related to the quality (e.g., defect density).
Software Reliability [STD 8739.8]	The discipline of software assurance that (1) defines the requirements for software controlled system fault/failure detection, isolation, and recovery; (2) reviews the software development processes and products for software error prevention and/or reduced functionality states; and (3) defines the process for measuring and analyzing defects and defines/derives the reliability and maintainability factors.
Software Safety [STD 8719.13]	The aspects of software engineering and software assurance that provide a systematic approach to identifying, analyzing, tracking, mitigating, and controlling hazards and hazardous functions of a system where software may contribute either to the hazard or to its mitigation or control, to ensure safe operation of the system.
Software Safety [STD 8739.8]	The discipline of software assurance that is a systematic approach to identifying, analyzing, tracking, mitigating, and controlling software hazards and hazardous functions (data and commands) to ensure safe operation within a system.
Software System Structure [STD 8739.9]	The specific organization of a software system's components for the purpose of accomplishing an operational capability.

Term [Citing Document(s)]	Definitions
Software-Related Event/Item [STD 8719.13]	Event (e.g., hazard, fault, failure mode or failure) or item (e.g., hazard control, requirement, function or mitigation) whose outcome or desired outcome is inextricably or causally linked to software functionality (i.e., without software, the event or the desired end-state as expressed by the item would not be possible).
Solar flux unit (sfu) [STD 8719.14]	Equal to 104 janskys measured at a wavelength of 10.7 cm.
Solder [STD 8739.4]	A nonferrous, fusible metallic alloy used to join metallic surfaces.
Solder Sleeve [STD 8739.4]	A heat-shrinkable solder termination device with meltable sealing preforms at ends.
Soldering [STD 8739.4]	The process of joining clean metallic surfaces through the use of solder without direct fusion of the base metals.
Solid Propellant [STD 8719.12]	Solid compositions used for propelling projectiles and rockets and to generate gases for powering auxiliary devices. Contains both fuels and oxidizers, and various other chemicals used as burning rate controllers and binders.
Solvent [STD 8739.1]	A non-reactive liquid substance that is capable of dissolving another substance (IPC T-50). Other terms used are: chemistry, cleaner (of any type), cleaning solution, cleaning solvent, detergent, saponifier, etc.
Source Code [STD 8739.9]	The collection of executable statements and commentary that implements detailed software design.
Source Control Drawing (SCD) [STD 8739.10]	A drawing that provides an engineering description (including configuration, part number, marking, reliability, environmental, and functional/performance characteristics), qualification requirements, and acceptance criteria for commercial items or vendor developed items procurable from a specialized segment of industry that provides for application critical or unique characteristics.
Soxhlet Extraction [STD 8739.5]	A process similar to distillation used to separate materials. Uses relative to this standard include removing oils, resins, or other contaminants from cotton swabs or wipes.
Space Debris [STD 8719.14]	General class of debris, including both meteoroids and orbital debris.
Space Structures [STD 8719.14]	Spacecraft and launch vehicle orbital stages. This includes all components contained within the object such as instruments and fuel.
Space System [NPR 8705.2]	The collection of all space-based and ground-based systems (encompassing hardware and software) used to conduct space missions or support activity in space, including, but not limited to, the crewed space system, space-based communication and navigation systems, launch systems, and mission/launch control. Also, referred to as "system" in the technical requirements.

Term [Citing Document(s)]	Definitions
Space Wing Commander [STD 8719.24 Annex]	in this document, the term Space Wing Commander refers exclusively to the commanders of the 30th Space Wing and the 45th Space Wing; the term Range Commander refers to the commander of the Eastern or Western Range in accordance with Department of Defense Directive 3200.11 and is the same individual as the Space Wing Commander; the terms Range Commander and Spacelift Commander refer to tasks or functions performed by the Space Wing Commander; see AFSPCI 10-1202, Crew Force Management, for further information.
Spacecraft [NPR 8621.1]	A habitable vehicle or device including, but not limited to, orbiters, capsules, modules, landers, transfer vehicles, rovers, Extravehicularactivity suits, and habitats, designed for travel or operation outside Earth's atmosphere.
Spacecraft [STD 8739.4]	Devices, manned or unmanned, which are designed to be placed into a suborbital trajectory, an orbit about the earth, or into a trajectory to another celestial body.
Spacecraft [STD 8719.14]	This includes all components contained within a space borne payload such as instruments and fuel.
Spares [STD 8729.1]	Maintenance replacements for parts, components, or assemblies in deployed items of equipment.
Special Use Airspace [STD 8719.25]	Airspace wherein activities must be confined because of their nature, or wherein limitations are imposed on aircraft operations that are not a part of those activities, or both. Warning areas, military operations areas, alert areas, and controlled firing areas are nonregulatory special use airspace. (Special Use Airspace is designated by the FAA.)
Specialized Fasteners [STD 6008]	Fasteners that fall under categories such as custom-designed and manufactured fasteners; pyrotechnic fasteners; non-metallic fasteners or commercial fasteners such as eyebolts, clevises, hooks, wire rope, turnbuckles, and continuous threaded rods; as well as those not otherwise specified.
Specification [STD 8739.9]	A document that specifies, in a complete, precise, verifiable manner, the requirements, design, behavior or other characteristics of a system or component, and, often the procedures for determining whether these procedures have been satisfied. (see IEEE Std 610.12-1990)
Specimen, Conformal Coating Test [STD 8739.1]	A spare PCB or similar substrate that is coated with the same material and process as used for the mission hardware that is used for evaluation of the quality of the conformal coating thickness.
Specimen, Materials Mix Test [STD 8739.1]	A test sample that meets the requirements in ASTM D2240, that is used to evaluate polymer mix quality using a hardness test. The sample of cured material is at least 6mm (0.24 in) thick by 24 mm (0.96 in) in diameter.
Splice [STD 8739.4]	The joining of two or more conductors to each other.
Splice [STD 8739.5]	An interconnection method for joining the ends of two optical fibers in a permanent or semi-permanent fashion.

Term [Citing Document(s)]	Definitions
Splice Enclosure [STD 8739.5]	A device surrounding the spliced area of an optical fiber used to protect the splice from physical damage.
Splice Tray [STD 8739.5]	A container used to organize and protect spliced fibers.
Splice, Chemical Splice [STD 8739.5]	A permanent joint made with an adhesive such as UV-cured polymer or epoxy.
Splice, Fusion Splice [STD 8739.5]	A splice accomplished by the application of localized heat sufficient to fuse or melt the ends of two lengths of optical fiber, forming a continuous single optical fiber.
Splice, Mechanical [STD 8739.5]	A fiber splice accomplished by fixtures or materials, rather than by thermal fusion.
Squeeze-out [STD 8739.1]	The resin and/or reinforcement that is visible at the edges of a bond.
Stabilized [STD 8719.14]	When the spacecraft maintains its orientation along one or more axes.
Staking [STD 8739.1]	The process of bonding and securing components or parts to PCBs and electronic assemblies by means of an adhesive material, with the intention to provide additional mechanical support.
Staking Material [STD 8739.1]	An electrically nonconductive adhesive material used for additional support.
Standard Operating Procedure (SOP) [STD 8719.24 Annex]	a procedure prepared for operation of a facility or performance of a task on a routine basis.
Standby LDE [STD 8719.9]	LDE not in regular service but used occasionally or intermittently as required. A lifting device or equipment that has not been used for a period of 1 month or more but less than 12 months is considered to be used intermittently/occasionally.
Static Electricity [STD 8719.12]	An accumulation of electrical charge on a conductive or dielectric material. Unbounded accumulation of electrical charge can result in high levels of potential difference leading to ESD events.
Static Test Stand [STD 8719.12]	Locations on which liquid propellant engines or solid propellant motors are tested in place.
Storage Compatibility [STD 8719.12]	A relationship between different items of explosives and other dangerous materials whose characteristics are such that a quantity of two or more of the items stored or transported together is no more hazardous than a comparable quantity of any one of the items stored alone.
Storage Magazine [STD 8719.12]	A structure designed or specifically designated for the long-term storage of explosives or ammunition.
Strain Relief	A connector device that prevents the disturbance of the contact and cable terminations.

Term [Citing Document(s)]	Definitions
[STD 8739.4]	
Stranded Conductor [STD 8739.4]	A conductor composed of a group of smaller wires.
Strength Member [STD 8739.5]	That part of a fiber optic cable composed of kevlar aramid yarn, steel strands, or fiberglass filaments included to increase the tensile strength of the cable, and in some applications, to support the weight of the cable.
stress intensity factor [STD 8719.24 Annex]	a parameter that characterizes the stress-strain behavior at the tip of a crack contained in a linear elastic, homogeneous, and isotropic body.
Stress Relief [STD 8739.1]	The formed portion of a conductor whose geometry minimizes stress on the mechanically clamped terminations.
Stress Relief [STD 8739.4]	The formed portion of a conductor that provides sufficient length to minimize stress between terminations.
Stress Screening [STD 8729.1]	The process of applying mechanical, electrical, or thermal stresses to an equipment item for the purpose of precipitating latent part and workmanship defects to early failure.
stress-corrosion cracking [STD 8719.24 Annex]	a mechanical-environmental induced failure process in which sustained tensile stress and chemical attack combine to initiate and propagate a crack or a crack-like flow in a metal part.
Strip [STD 8739.4]	To remove insulation from a conductor.
structural component [STD 8719.24 Annex]	a component such as a bolt, lug, hook, shackle, pin, rivet, or weld in a piece of material handling equipment.
Structural Sling [STD 8719.9]	A term sometimes used for rigid or semi-rigid lifting devices such as spreader bars or lifting beams that now are included in the general category of below-the-hook lifting devices.
structural sling [STD 8719.24 Annex]	a rigid or semi-rigid fixture that is used between the actual object being lifted and hoisting device hook (e.g., spreader bars, equalizer bars, and lifting beams).
Subject Matter Expert [STD 8709.20]	Person recognized as an expert in the technical area under review.
Substantial Damage [NPR 8621.1]	Damage or failure adversely affecting structural strength, performance, or flight characteristics of an aircraft, which would normally require major repair or component replacement. Engine failure or damage limited to an engine if only one engine fails or is damaged, bent fairings or cowling, dented skin, small punctured holes in the skin or fabric, ground damage to rotor or propeller blades, and damage to landing gear, wheels, tires, flaps, engine accessories, brakes, or wingtips are not considered substantial damage (49 CFR pt. 830).
Substrate [STD 8739.1]	That surface upon which an adhesive is spread for any purpose, such as coating; a broader term than "adherent."

Term [Citing Document(s)]	Definitions
Subsystem [NPR 8705.2]	A secondary or subordinate system within a system (such as the crewed space system) that performs a specific function or functions. Examples include electrical power, guidance and navigation, attitude control, telemetry, thermal control, propulsion, structures subsystems. A subsystem may consist of several components (hardware and software) and may include interconnection items such as cables or tubing and the support structure to which they are mounted.
Subsystem [STD 8729.1]	A grouping of items satisfying a logical group of functions within a system.
Supplier [STD 6008]	A fastener manufacturer or distributor.
Supplier [STD 8709.20]	The organization that applies the CM [configuration management] discipline. The supplier may be a contractor, academia, or the Government. The supplier may be the design agency involved in production of a product, or be limited to producing documentation. Note: The role of "contractor" is not defined in this document and is assumed to be included within the role of "supplier" (from NASA-STD 0005).
Supplier [STD 8739.4]	In-house NASA, NASA contractors, and subtier contractors.
Supplier [STD 8739.6]	Any entity who is manufacturing hardware in accordance with the requirements herein, including NASA Centers and NASA contractors.
Supplier [STD 8729.1]	Any organization, which provides a product or service to a customer. By this definition, suppliers may include vendors, subcontractors, contractors, flight programs/projects, and the NASA organization supplying science data to a principal investigator. (In contrast, the classical definition of a supplier is: a subcontractor, at any tier, performing contract services or producing the contract articles for a contractor.).
Support Equipment [STD 8729.1]	Equipment required to maintain systems in effective operating condition in its intended environment, including all equipment required to maintain and operate the system and related software.
surface inspection [STD 8719.24 Annex]	a nondestructive examination method, other than visual, used for detection of surface and near surface discontinuities.
Surface Nondestructive Testing [STD 8719.9]	Test and inspection methods used to examine the surface of equipment/materials (e.g., magnetic particle and liquid penetrant).
Surveillance [STD 8739.8]	The continuous monitoring and status of an entity and analysis of records to ensure that specified requirements are being met.
Surveillance Inspection [STD 8719.12]	Periodic visual inspection of explosive stock to determine serviceability and/or storage conditions.
Suspect Counterfeit [NPR 8735.1]	An item for which credible evidence (including but not limited to, visual inspection or testing) provides reasonable doubt that the item is authentic.

Term [Citing Document(s)]	Definitions
Sustainability [STD 8729.1]	The ability to maintain the necessary level and duration of logistics support to achieve mission objectives.
Sustainment [STD 8729.1]	The provision of logistics and personnel services required to maintain and prolong operations until successful mission accomplishment.
System [STD 8739.9]	The combination of elements that function together to produce the capability required to meet a need. The elements include hardware, software, equipment, facilities, personnel, processes, and procedures needed for this purpose.
System [STD 8729.1]	[1] The combination of elements that function together to produce the capability to meet a need. The elements include all hardware, software, equipment, facilities, personnel, processes, and procedures needed for this purpose. [2] The end product (which performs operational functions) and enabling products (which provide life-cycle support services to the operational end products) that make up a system.
system hazard [STD 8719.24 Annex]	a hazard associated with a hardware system and that generally exists even when no operation is occurring; system hazards that may be found at a launch site include, but are not limited to, explosives and other ordnance, solid and liquid propellants, toxic and radioactive materials, asphyxiants, cryogens, and high pressure.
System Safety [NPR 8715.3] [NPR 8715.7] [STD 8719.13]	Application of engineering and management principles, criteria, and techniques to optimize safety and reduce risks within the constraints of operational effectiveness, time, and cost throughout all phases of the system life cycle.
System Safety Manager [NPR 8715.3]	A designated management person who, qualified by training and/or experience, is responsible to ensure accomplishment of system safety tasks.
System Safety Plan [NPR 8715.7]	A written plan defining the approach to accomplish the project safety activities, including safety management, identification of safety tasks, roles and responsibilities, and the coordination and communication with project/systems engineers and approving authorities. It is also known as the System Safety Technical Plan as defined in NPR 8715.3, and the Systems Safety Program Plan defined in Air Force Space Command Manual 91-710, Range Safety User Requirements Manual, Volume III, Chapter 4.
Tailoring [STD 8719.25]	The process where the authorities responsible for range safety requirements and a range user review each requirement and jointly document whether the requirement is applicable to the range user's planned operations and, if it is applicable, document whether the range user will meet the requirement as written or achieve an equivalent level of safety through an acceptable alternative. Tailoring includes Equivalent Level of Safety determinations. Tailoring does not include the approval of Range Safety Waivers, which are addressed by a separate process.
Tailoring [NPR 8715.7]	The process of assessing the applicability of requirements and evaluating the project's potential implementation in order to generate a set of specific requirements for the project.
Tailoring [STD 8709.20]	The process used to refine or modify an applicable requirement by the implementer of the requirement. If the revised requirement meets/exceeds the original requirement, and has no increase in risk from that of the original requirement, then it may be accepted/implemented by appropriate local authority; otherwise a waiver/deviation may be required.

Term [Citing Document(s)]	Definitions
Tailoring [STD 8719.13]	The process used to refine or modify an applicable requirement by the implementer of the requirement. If the revised requirement meets/exceeds the original requirement, and has no increase in risk from that of the original requirement, then it may be accepted/implemented by appropriate local authority; otherwise a waiver/deviation may be required. This definition of "tailoring" is for use within the application of the SMA TA as defined by the Chief, Safety and Mission Assurance.
Tailoring [STD 8739.9]	[1] The process of assessing the applicability of requirements and evaluating the project's potential implementation in order to generate a set of specific requirements for the project. [2] The process used to refine or modify an applicable requirement by the implementer of the requirement. If the revised requirement meets/exceeds the original requirement, and has no increase in risk from that of the original requirement, then it may be accepted/implemented by appropriate local authority; otherwise a waiver/deviation may be required.
Tailoring [STD 8729.1]	The process used to adjust or seek relief from a prescribed requirement to accommodate the needs of a specific task or activity (e.g., program or project). The tailoring process results in the generation of deviations and waivers depending on the timing of the request (Source: NPR 7120.5, NASA Space Flight Program and Project Management Requirements).
Tang (Connector Backshell) [STD 8739.4]	A backshell tang is a tapering metal projection (straight, 45°, or 90° to the axis of the connector) designed to accommodate cable-tie attachments. The cable-ties grip and hold harness wires exiting from the connector, thus providing stress relief for the wires.
Task [STD 8729.1]	A function to be performed. In contract proposals, a unit of work that is sufficiently well defined so that, within the context of related tasks, readiness criteria, completion criteria, cost and schedule can all be determined.
Technical Authority [NPR 8705.2]	The individuals who provide independent oversight of programs and projects in support of safety and mission success, who have formally delegated authority traceable to the Administrator, and are funded independent of Programmatic Authority. (Source: paraphrased from NPD 1000.0)
Technical Authority [STD 6008]	The agency or organization that is responsible for the technical details of a particular design and the resolution of any associated technical issues.
Technical Authority [STD 8739.10]	Individuals at different levels of responsibility who maintain independent authority to ensure that proper technical standards are utilized.
Technical Waiver [STD 8719.17]	Single or case-by-case waiver from this standard.
telemetry [STD 8719.24 Annex]	vehicle systems measurements made available to ground based users via S-band downlinks.
Temporary Holding Area [STD 8719.12]	Designated areas for temporarily parking explosive laden transport trucks/railcars. QD and compatibility requirements apply.

Term [Citing Document(s)]	Definitions
Test [NPR 8621.1] [STD 8729.1]	A procedure for critical evaluation; a means of determining the presence, quality, or truth of something; a trial. In engineering, a method of determining performance by exercising or operating a system or item using instrumentation or special test equipment that is not an integral part of the item being tested.
Test Article PVS [STD 8719.17]	A PVS object(s) being tested for the sole purpose of obtaining data (other than integrity data) on the object(s).
Test Flight [NPR 8705.2]	A flight or mission dedicated primarily to test objectives. Flight tests can include scaled test articles, uncrewed flights, and crewed flights.
Test Plan [STD 8739.9]	A document prescribing the approach to be taken for intended testing activities. The plan typically identifies the items to be tested, the testing to be performed, test schedules, personnel requirements, reporting requirements, evaluation criteria, the level of acceptable risk, and any risk requiring contingency planning.
Test Procedure [STD 8739.9]	The detailed instructions for the setup, operation, and evaluation of results for a given test. A set of associated procedures is often combined to form a test procedure document.
Test Specific PVS [STD 8719.17]	PVS used to perform limited testing of a specific test article. PVS used on a permanent or repeated basis, or built up of components used repeatedly for testing different hardware or configurations are not part of this category.
Test Specimen [STD 8739.1]	See Specimen.
Testability [STD 8729.1]	A design characteristic that permits timely and cost-effective determination of the status (operable, inoperable or degraded) of a system or subsystem with a high level of confidence. Testability attempts to quantify those attributes of system design that facilitate detection and isolation of faults that affect system performance.
testing laboratory (nationally recognized) [STD 8719.24 Annex]	laboratories such as Underwriters Laboratories, Inc., or Factory Mutual Engineering Corporation, that use nationally recognized testing standards and provide bench mark(s) to certified products as evidence of successful testing.
Threshold [NPR 8000.4]	A level for a performance measure or a risk metric whose exceedance "triggers" management processes to rectify performance shortfalls.
threshold limit value [STD 8719.24 Annex]	time weighted average concentrations that must not be exceeded during any 8-hour work shift of a 40-hour work week as determined the American Conference of Governmental Industrial Hygienists.
Tightly Coupled Program [STD 8709.20]	Programs having multiple projects that execute portions of a mission or missions. No single project is capable of implementing a complete mission. Typically, multiple NASA Centers contribute to the program. Individual projects may be managed at different Centers. The program may also include other Agency or international partner contributions. (This definition is from NASA Memo 7120-81, paragraph 2.1.4.d, which updated NPR 7120.5D)

Term [Citing Document(s)]	Definitions
Timeline [NPR 8621.1]	Events and conditions preceding and following a mishap supported by facts and arranged in chronological order.
Tines [STD 8739.4]	Tines are the members of a contact retention system that capture or "lock" removable crimp contacts into the contact cavities.
to safe [STD 8719.24 Annex]	to bring to a safe condition.
Total Ionizing Dose (TID) [STD 8739.10]	The cummulative energy deposited in a material causing a long-term degradation of electronics. Typical effects include parametric failures or variations in device parameters such as leakage current, threshold voltage, etc., or functional failures.
toxic hazard zone [STD 8719.24 Annex]	a generic term that describes an area in which predicted concentration of propellant or toxic byproduct vapors or aerosols may exceed acceptable tier levels; predictions are based on an analysis of potential source strength, applicable exposure limit, and prevailing meteorological conditions; toxic hazard zones are plotted for potential, planned, and unplanned propellant releases, and launch operations.
Traceability [STD 8739.9]	[1] The degree to which a relationship can be established between two or more products of the development process, especially products having a predecessor successor or master-subordinate relationship to one another; for example, the degree to which the requirements and design of a given software component match (see IEEE Std 610.12-1990). [2] The characteristic of a system that allows identification and control of relationships between requirements, software components, data, and documentation at different levels in the system hierarchy.
Traceability [STD 8739.10]	The ability to verify the history, location, or application of an item by means of documented recorded identification.
Traceability Code [STD 8739.1]	The code uniquely identifying the production lot by the manufacturer, equivalent to batch code, lot code, or date code.
Transient [STD 8719.12]	A person with official business on a production line or operation but who is not routinely assigned to a specific limited location. Typically, transients are roving supervisors, quality assurance, safety personnel, or maintenance personnel. Official visitors are considered transients.
Transmissivity [STD 8739.1]	The fractional quantity of incident radiation transmitted by matter.
Two-Block [STD 8719.9]	A condition in which the lower load block or hook assembly comes into contact with the upper load block, hoist/trolley structure, or boom point sheave assembly.
Type A Mishap [NPR 8621.1]	A mishap resulting in one or more of the following: a. Occupational injury or illness resulting in a fatality or a permanent total disability. b. Total direct cost of mission failure and property damage of $2,000,000 or more. c. Crewed aircraft hull loss. d. Unexpected aircraft departure from controlled flight for all aircraft except when departure from controlled flight has been pre-briefed (e.g., upset recovery training, high AOA envelope testing, aerobatics, or Out of Controlled Flight for training) or mitigated through the flight test process inherent at each Center.

Term [Citing Document(s)]	Definitions
Type B Mishap [NPR 8621.1]	A mishap causing an occupational injury or illness resulting in permanent partial disability; hospitalization for inpatient care of three or more people within 30 workdays of the mishap; or a total direct cost of mission failure and property damage of at least $500,000, but less than $2,000,000.
Type C Mishap [NPR 8621.1]	A mishap resulting in a nonfatal OSHA-recordable occupational injury or illness causing days away from work, restricted duty, or transfer to another job beyond the day or shift on which the mishap occurred; hospitalization for inpatient care of one or two people within 30 workdays of the mishap; or a total direct cost of mission failure and property damage of at least $50,000 but less than $500,000.
Type D Mishap [NPR 8621.1]	A mishap resulting in a nonfatal OSHA-recordable occupational injury or illness that does not meet the definition of a Type C mishap or a total direct cost of mission failure and property damage of at least $20,000, but less than $50,000.
ultimate load [STD 8719.24 Annex]	the product of the limit load and the design ultimate load factor. It is the load that the structure must withstand without rupture or collapse in the expected operating environment.
ultimate strength [STD 8719.24 Annex]	the maximum stress developed by the material before rupture, based on the original area, in tension, compression, or shear; see Modern Steels and Their Properties, Carbon and Alloy Steel Bars and Rods in References.
Ultraviolet (UV) [STD 8739.5]	Optical radiation for which the wavelengths are shorter than those for visible radiation that is approximately between 1nm and 400nm.
Uncertainty [NPR 8000.4]	An imperfect state of knowledge or a variability resulting from a variety of factors including, but not limited to, lack of knowledge, applicability of information, physical variation, randomness or stochastic behavior, indeterminacy, judgment, and approximation.
Undesired Outcome [NPR 8621.1]	An event or result that is unwanted and different from the desired and expected outcome. For mishap investigation, an undesired outcome should describe the loss that determined the mishap classification (i.e., property damage, mission failure, fatality, permanent disability, lost-time case, or first-aid case).
Unmanned Aerial Vehicle (UAV) [NPR 8715.5]	A vehicle without a pilot on board that is controlled autonomously by an onboard control and guidance system or is controlled from a monitoring station outside of or remote from the UAV vehicle. A UAV is defined as an aircraft by the FAA.
Unmanned Aerial Vehicle (UAV) [STD 8719.25]	A vehicle that is controlled remotely or that is autonomous and operates at speeds ranging from subsonic to hypersonic in a manner consistent with a "conventional" aircraft. A UAV may be launched from the ground or dropped from other aerial vehicles, subscale flight test vehicles, or lifting bodies. A UAV may also be referred to using a different name such as Uninhabited Air Vehicle, Unmanned Aircraft, Drone, Remotely Piloted Aircraft, Remotely Operated Aircraft, or Remotely Piloted Vehicle.
Unmanned Aircraft System (UAS) [STD 8719.25]	A UAS includes a UAV or similar vehicle and all the associated support equipment, control station, data links, telemetry, communications and navigation equipment necessary to operate the vehicle. A UAS can be operated via a remotely located, manually operated flight control system or ground control system.

Term [Citing Document(s)]	Definitions
Unmanned Aircraft Systems (UAS) [NPR 8715.5]	A UAS includes an Unmanned Aerial Vehicle (UAV) or similar vehicle and all the associated support equipment, control station, data links, telemetry, and communications and navigation equipment necessary to operate the vehicle. UAS can be operated via a remotely located, manually operated flight control system or ground control system.
Unserviceable Explosive [STD 8719.12]	An explosive material or device which is not currently qualified for its original intended purpose, but may be inspected and returned to service for its original intended purpose or may with proper analysis be used for another purpose, such as for demonstrations, research, testing, training, etc. Unserviceable materials and devices are not considered or managed as hazardous waste until the material(s) or device(s) are declared to be waste and are earmarked for disposal.
Usability Testing [NPR 8705.2]	Evaluation by people using the system (hardware or software) in a realistic situation to determine how well it can be used for its intended purpose (e.g., how well people can manipulate parts or controls, receive feedback, and interpret feedback) to identify potential human errors and areas for design improvement.
Usage [NPR 8735.1]	The item presented in the NASA Advisory, GIDEP Notice, or other released document is present on the system being evaluated.
Utilities [STD 8719.12]	Those services such as water, air, steam, sewage, telephone, and electricity necessary to the operation of an establishment.
Vacuum System [NPR 8715.3]	An assembly of components under vacuum, including vessels, piping, valves, relief devices, pumps, expansion joints, gages, and others.
Validation [NPR 8705.2]	Proof that the product accomplishes the intended purpose. May be determined by a combination of test, analysis, and demonstration.
Validation [NPR 8715.3]	(1) An evaluation technique to support or corroborate safety requirements to ensure necessary functions are complete and traceable; or (2) the process of evaluating software at the end of the software development process to ensure compliance with software requirements.
Validation [STD 8739.8]	Confirmation by examination and provision of objective evidence that the particular requirements for a specific intended use are fulfilled. [ISO/IEC 12207, Software life cycle processes] In other words, validation ensures that "you built the right thing."
Validation [STD 8739.9]	Confirmation that the product, as provided (or as it will be provided), fulfills its intended use. In other words, validation ensures that "you built the right thing." (see SEI-CMMI)
Validation [STD 8729.1]	To establish the soundness of, or to corroborate. As a process, validation answers, "Are we building the right system?" Validation testing of products is performed to ensure that each reflects an accurate interpretation and execution of requirements and meets a level of functionality and performance that is acceptable to the user or customer.
Variance [NPR 8715.3]	An authorization for temporary relief in advance from a specific requirement and is requested during the formulation/planning/design stages of a program/project operation to address expected situations.

Term [Citing Document(s)]	Definitions
vehicle [STD 8719.24 Annex]	launch vehicle and/or payload.
Vendor Hi-Rel [STD 8739.10]	A term used to describe parts that have been screened and qualified to requirements that have been enhanced from the manufacturer's normal flow, as determined solely by the manufacturer and offered as high reliability parts.
Verification [NPR 8705.2]	Proof of compliance with specifications. May be determined by a combination of test, analysis, demonstration, and inspection.
Verification [STD 8739.8]	Confirmation by examination and provision of objective evidence that specified requirements have been fulfilled. [ISO/IEC 12207, Software life cycle processes] In other words, verification ensures that "you built it right."
Verification [STD 8739.9]	Confirmation that work products properly reflect the requirements specified for them. In other words, verification ensures that "you built it right." (see SEI-CMMI)
Verification [STD 8729.1]	The task of determining whether a system or item meets the requirements established for it. As a process, verification answers, "Are we building the system right?"
Verification (Software) [NPR 8715.3]	(1) The process of determining whether the products of a given phase of the software development cycle fulfill the requirements established during the previous phase (see also validation); or (2) formal proof of program correctness; or (3) the act of reviewing, inspecting, testing, checking, auditing, or otherwise establishing and documenting whether items, processes, services, or documents conform to specified requirements.
Verification Plan [NPR 8705.2]	A formal document listing the specific technical process to be used to show compliance with each requirement.
Vernal Equinox [STD 8719.14]	The direction of the Sun in space when it passes from the southern hemisphere to the northern hemisphere (on March 20 or 21) and appears to cross the Earth's equator. The vernal equinox is the reference point for measuring angular distance along the Earth's equatorial plane (right ascension) and one of two angles usually used to locate objects in orbit (the other being declination).
Viscosity [STD 8739.1]	A measure of the resistance of a material to flow under stress.
visible damage [STD 8719.24 Annex]	for composite pressure vessels; Anomalies that are visible to the naked eye under not less than 15-foot candles at a distance no greater than 24 inches and no less than a 30 degree angle. Lighting up to 50-foot candles may be used for the detection or study of small anomalies.
Volumetric Nondestructive Testing [STD 8719.9]	Test and inspection methods used to examine the interior of equipment/materials (e.g., ultrasonic and radiographic).

Term [Citing Document(s)]	Definitions
Voluntary Consensus Standards (VCS) [STD 8719.9]	Industry standards used by NASA for LDE design, operations, maintenance, and inspections, including American Gear Manufacturers Association (AGMA), American Society of Mechanical Engineers (ASME), Deutsches Institut für Normung (DIN), American National Standards Institute (ANSI), and Scaffold and Access Industry Association (SAIA).
Waiver [NPR 8705.2]	A documented authorization releasing a program or project from meeting a requirement after the requirement is put under configuration control at the level the requirement will be implemented (source NPD 7120.4), where a certain level of risk has been documented and accepted.
Waiver [NPR 8715.3]	A variance that authorizes departure from a specific safety requirement where a certain level of risk has been documented and accepted.
Waiver [NPR 8715.7]	A written authorization granting relief from an applicable requirement and documenting the acceptance of any associated risk. For NASA ELV payload projects, waivers typically are approved for a single mission and have a specific duration. However, a waiver identified early in the design or specification/requirement review(s) may apply throughout the project or to multiple missions that use a common upper stage and/or a common spacecraft bus.
Waiver [STD 8709.20]	(1) A written authorization to depart from a specific directive requirement (from NPR 1400.1). (2) A documented authorization releasing a program or project from meeting a requirement after the requirement is put under configuration control at the level the requirement will be implemented This definition is from NASA Memo 7120-81, Appendix A, which updated NPR 7120.5D).
Waiver [STD 8719.12]	A documented authorization releasing a program or project from meeting a requirement after the requirement is put under configuration control at the level the requirement will be implemented.
Waiver [STD 8719.17]	As defined in NPR 8715.3.
Waiver [STD 8739.10]	A written authorization, granted after manufacture, to accept a CI that is found to depart from specified requirement(s) of the CI's current approved configuration for a specific number of units or a specified period of time.
waiver [STD 8719.24 Annex]	a variance that authorizes departure from a specific safety requirement where a certain level of risk has been documented and accepted; a designation used when, through an error in the manufacturing process or for other reasons, a hardware noncompliance is discovered after hardware production, or an operational noncompliance is discovered after operations have begun at the Eastern or Western ranges.
Waiver, Fleet [STD 8709.20]	A type of waiver which will apply to multiple products/situations that are using the same design or operation which may be outside of a single program/project or Center.

Term [Citing Document(s)]	Definitions
Walkthrough [STD 8739.9]	A static analysis technique in which a designer or programmer leads members of the development team and other interested parties through a software product, and the participants ask questions and make comments about possible anomalies, violation of development standards, and other problems. (see IEEE Std. 1028-2008)
Western Range [STD 8719.24 Annex]	part of the National Launch Range facilities, operated by the 30th Space Wing, part of Air Force Space Command, and located at Vandenberg Air Force Base, California; the range includes the operational launch and base support facilities located at Vandenberg Air Force Base and those radar tracking sites and ground stations located on sites uprange and downrange along the Pacific Coast, including United States Navy facilities at Point Mugu.
White plague [STD 8739.10]	Reaction occuring when excess fluorine outgasses from fluoropolymer insulations combines with water in the form of humidity to create hydrofluoric acid, which reacts with any surrounding metal.
Wicking [STD 8739.4]	A flow of molten solder, flux, or cleaning solution by capillary action.
Will [STD 8709.20]	The verb "will" describes a fact, expectation, or premise of accomplishment by a Program/Project/Center (from NASA-STD 0005).
Winch [STD 8719.9]	A stationary motor-driven or hand-powered hoisting machine having a drum around which is wound a rope, chain, or web used for lifting and lowering a load (requirements in this standard do not apply to winches used for horizontal pulls).
Wire [STD 8739.4]	A single metallic conductor of solid, stranded, or tinsel construction, designed to carry currents in an electrical circuit. It may be bare or insulated.
Wire Dress [STD 8739.4]	The arrangement of wires and laced harnesses in an orderly manner.
Witness [NPR 8621.1]	A person who has information, evidence, or proof about a mishap and provides his or her knowledge of the facts to the investigating authority.
Witness Statement [NPR 8621.1]	A verbal or written statement from a witness of his or her account including a description of the sequence of events, facts, conditions, and causes of the mishap.
Work Breakdown Structure (WBS) [NPR 8735.1]	A product-oriented hierarchical division of the hardware, software, services, and data required to produce the program's or project's end product(s), structured according to the way the work will be performed and reflecting the way in which program/project costs and schedule, technical, and risk data are to be accumulated, summarized, and reported.
Work Product [STD 8739.9]	The output of a task. Formal work products are deliverable to the acquirer. Informal work products are necessary to an engineering task but not deliverable. A work product may be an input to a task.
Working Life [STD 8739.5]	The duration of time that an adhesive can be used for a process and still achieve the intended result, with the intended quality, as measured from the time it is mixed or is fully defrosted, in the case of frozen sub-batches.
yield factor of safety	see factor of safety, yield.

Term [Citing Document(s)]	Definitions
[STD 8719.24 Annex]	
yield point [STD 8719.24 Annex]	see yield strength.
yield strength [STD 8719.24 Annex]	the stress at which there is an appreciable increase in strain with no increase in stress; typically defined as the stress that will induce a specified permanent set (yield point, usually 0.2 percent strain offset); see Mechanics of Materials and Modern Steels and Their Properties, Carbon and Alloy Steel Bars and Rods in References.

www.ingramcontent.com/pod-product-compliance
Lightning Source LLC
Chambersburg PA
CBHW081728220526
45468CB00008B/2011